Remote Sensing
in
Hydrology

REMOTE SENSING APPLICATIONS

Series Editors

E. C. Barrett
Director, Remote Sensing Unit
University of Bristol

L. F. Curtis, OBE
Honorary Senior Research Fellow,
University of Bristol

In the same series:

Remote Sensing of Ice and Snow
Dorothy K. Hall and Jaroslav Martinec

Imaging Radar for Resources Surveys
J. William Trevett

Remote Sensing in Hydrology

E. T. ENGMAN
*Hydrological Sciences Branch, NASA,
Goddard Space Flight Center, Greenbelt,
Maryland, USA*

and

R. J. GURNEY
*NERC Unit for Themic Information Systems
University of Reading, Reading, UK*

CHAPMAN AND HALL

LONDON • NEW YORK • TOKYO • MELBOURNE • MADRAS

UK   Chapman and Hall, 2–6 Boundary Row, London SE1 8HN

USA   Van Nostrand Reinhold, 115 5th Avenue, New York NY10003

JAPAN   Chapman and Hall Japan, Thomson Publishing Japan, Hirakawacho
            Nemoto Building, 7F, 1-7-11 Hirakawa-cho, Chiyoda-ku, Tokyo 102

AUSTRALIA   Chapman and Hall Australia, Thomas Nelson Australia,
            102 Dodds Street, South Melbourne, Victoria 3205

INDIA   Chapman and Hall India, R. Seshadri, 32 Second Main Road, CIT East,
            Madras 600 035

First edition 1991

© 1991 E. T. Engman and R. J. Gurney
Softcover reprint of the hardcover 1st edition 1991

Typeset in 10/12pt Plantin by EJS Chemical Composition, Bath, Avon

ISBN 978-94-010-6670-9

British Library Cataloguing in Publication Data

Engman, E.T.
    Remote sensing in hydrology.
    1. Hydrology. Applications of remote sensing
    I. Title   II. Gurney, R.J.   III. Series
    551.48028

    ISBN 978-94-010-6670-9   ISBN 978-94-009-0407-1 (eBook)
    DOI 10.1007/978-94-009-0407-1

Library of Congress Cataloguing-in-Publication Data

available

To our families

Contents

Acknowledgements

Very few technical books can be written in a vacuum. In fact, it is the nature of technology that each new step or finding is built on the foundation provided by many others. Thus it is with this book. Our summary of what we think is the status of remote sensing in hydrology clearly represents the efforts of many individuals and institutions. To those many who cannot all be acknowledged by name we are extremely grateful.

There are other individuals and colleagues who actively contributed to our efforts. We would like to acknowledge especially Vince Salomonson for his general encouragement and Joyce Tippett for assembling our many indecipherable notes and drafts into a finished manuscript. We are equally indebted to Walt Blanchette of Tara, Inc. for redrawing the many figures that appear throughout the book.

There are also many fellow scientists that either obtained photos or loaned us figures and imagery from their printed papers. In this regard we owe a special thanks to Peggy O'Neill, Diane Evans, Darrell Williams, Al Chang, Al Rango, Jerry Ritchie, Rod Schofield, Moe Deutsch, Max Miller, Fawwaz Ulaby, H. Kaufmann, Gerry Schaber, Richard Stumpf, Siamak Khorram, Pat Chavez, Floyd Sabins, Mark Imhoff and B. J. Conway.

Lastly we would like to acknowledge that group of dedicated and patient colleagues that reviewed various chapters and parts of chapters for us. Our special thanks to Tom Jackson, Al Rango, Jerry Ritchies, Tom Schmugge, Victor van Katwijk, Locke Stuart, Bhaskar Choudhury and Bill Kustas.

Preface

Remote sensing has been used for maybe eighty years as a practical tool to aid mapping. For many years, most of the work has been very qualitative and involved mainly aerial photointerpretation. More quantitative measurements have gradually been introduced, as new instruments have been developed, and remote sensing has gradually been accepted more widely in the earth sciences as a useful technique. Generally, the use of quantitative data for earth sciences has occurred first in those areas such as atmospheric sciences, oceanography and solid-earth geophysics where few conventional data are available. Those subjects where many conventional data are available, as in hydrology, have been more conservative in introducing remote sensing techniques. However, the growth of hydrological sciences has now resulted in a much greater need for data in different areas including in remote areas where few if any data are available, and hence the demand for remotely sensed data has increased. There has also been a shift in emphasis from near-term water resources projects to hydrological sciences, which implies that more effort can be put in developing new models, which is necessary for using remotely sensed data most effectively.

The nature of remote sensing itself has changed over the past few decades. It has moved from a relatively qualitative 'art' relying on inference for information, to a quantitative 'science' capable of measuring system states in some cases. With respect to hydrologic applications, there are several unique aspects of remote sensing. Perhaps first and foremost is the ability to measure spatial information as opposed to point data from which most of our hydrologic concepts and models have been developed. Close behind this is the ability to measure actual state variables over a drainage basin; variables such as soil moisture, surface temperature, snow water equivalent, the presence of liquid water in the snow and whether or not a soil may be frozen. Another important facet of remote sensing is limited to satellite sensors, and this is the potential ability to assemble long term temporal data sets—and by long term we mean years to decades.

The table of contents of this book looks much like any conventional hydrology textbook, except for the addition of a chapter on soil moisture. This chapter demonstrates the unique nature of the remote sensing science. Soil moisture is an environmental descriptor and a basin state variable that hydrologists have not been able to adequately account for with traditional point measurements. The science of remote sensing has given the hydrologist a capability of measuring this very important part of the cycle. The ability to measure the state of the surface over large areas, whether soil moisture or snow or the surface energy balance, really distinguishes remote sensing from conventional data collection and allows different types of models to be developed, and different scientific questions to be addressed.

As the number of applications of remotely sensed data have increased, so there has been a need to summarize the advances that have been made. We hope that this book at least partly meets this need. The subject of this book is changing rapidly. The various international initiatives related to global change and climate are leading to rapid progress both in modelling and in data collection. We hope that despite this rapid progress, that this volume will be useful for those interested in this field, and await further progress with anticipation and great interest.

Greenbelt,   E. T. Engman
Maryland   R. J. Gurney

1

Hydrologic cycle

1.1 INTRODUCTION

Water is one of the most prevalent substances in the Earth–atmosphere system and it is one of the most fundamental to the existence of life as we know it. Vernadskii (1960) has eloquently stated:

> Water occupies a unique position in the history of our planet. No other natural substance can be compared with it insofar as its influence on the course of the most basic geological process is concerned. All substances on this Earth — whether mineral, rock, or living body — contain some water. All matter on the Earth is affected by water in some way and is either impregnated with it or surrounded by it.

Certainly the existence of abundant water in all three phases (solid, liquid and gaseous) in the environment provides one of the most distinctive characteristics of the Earth compared with other planets of the solar system.

Understanding the general laws governing the distribution of water over the Earth and the collection of basic data on the water balance of the sea and river basins, continents and the planet as a whole is essential to the rational use and protection of water resources and for understanding the evolution of our planet. In short, the study of the movement of water on the Earth is of very practical importance for life on the planet. Historically, the existence and expansion of civilizations have been controlled to a great degree by the abundance, or shortage, of water. Likewise, hydrology has been directed for the most part to the management and control of water for the benefit of the human race. Because of this, hydrology and hydrology books have a distinctly engineering flavour, as opposed to a more scientific direction. A result of this rather practical, problem-solving perspective to hydrology is that the problems of drinking water, irrigation, floods and waste disposal have all been defined within relatively local-scale perspectives.

Recent questions concerning hydrology have begun to expand out from the local or regional scales to consider the global situation. The pressures that a growing world population and intense industrialization have put on the resources of this planet are unique to this century. Problems that once were local or regional are now continental and global. No longer are single-purpose engineering approaches and solutions satisfactory. Water-related problems and solutions directly and indirectly affect many aspects of our lives and our use of other natural resources. In order to understand the complex interactions that affect our water resources, it is no longer sufficient to examine one phase of a hydrologic problem or solution; we must begin to look at the large-scale ramifications. In the twentieth century, hydrology has, out of necessity, expanded its framework from the basin to the globe. Remote sensing offers a unique approach for studying hydrology across these disparate scales.

1.2   HYDROLOGIC CYCLE

Because of this expanded scope in hydrology, it is necessary to examine the hydrologic cycle from differing perspectives. Before doing so, however, a definition may be helpful: the hydrologic cycle is the endless recirculation of water from water vapour to precipitation, streamflow, lakes and oceans, and returning to water vapour through evaporation and transpiration. Figure 1.1 illustrates this cycle schematically and Figure 1.2 shows a slightly more complex engineering view. How we study the hydrologic cycle depends upon the scale of reference. From a global perspective the hydrologic cycle is complete, but within a catchment the cycle does not close, although over a long period of time it does balance itself with regard to inputs and outputs. Because of these differences it may be useful to examine the hydrologic cycle from these two perspectives.

Global hydrologic cycle

From a global perspective, the equivalent depth of about 1 metre of precipitation falls on the surface of the Earth and is evaporated each year (Dooge 1973). In terms of the overall water balance of the Earth (shown in Table 1.1) and from the perspective of the cycle shown in Figures 1.1 and 1.2 showing the fluxes between the various storage parameters in the hydrologic cycle, we can begin to appreciate better the complexities and magnitudes of the task facing scientists if they wish to observe, model and understand the hydrologic cycle and its variations over time and space. The problem is further

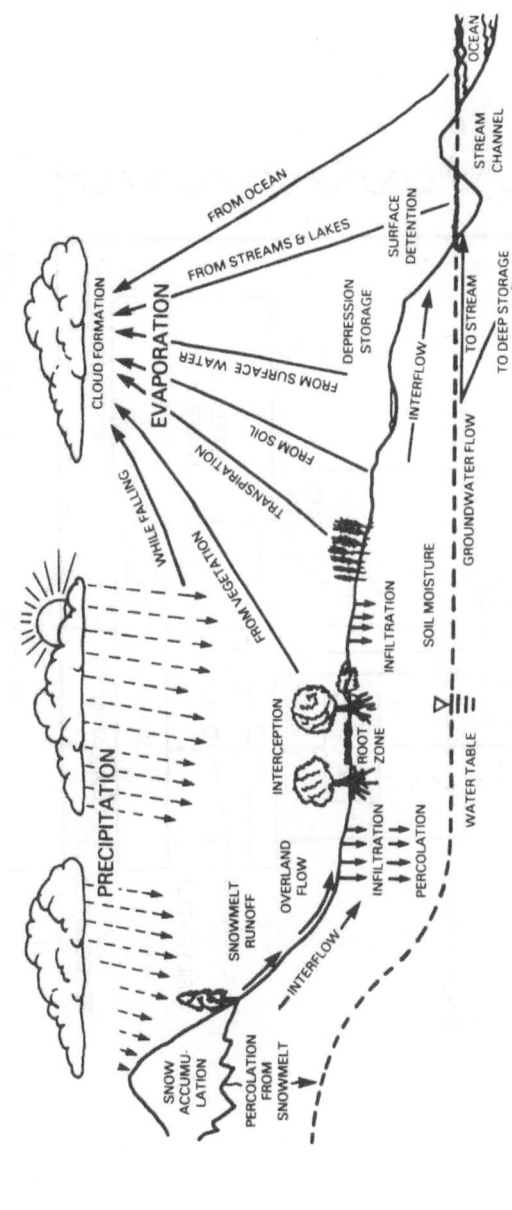

Figure 1.1 A schematic representing processes in the global hydrologic cycle (after NASA 1984a).

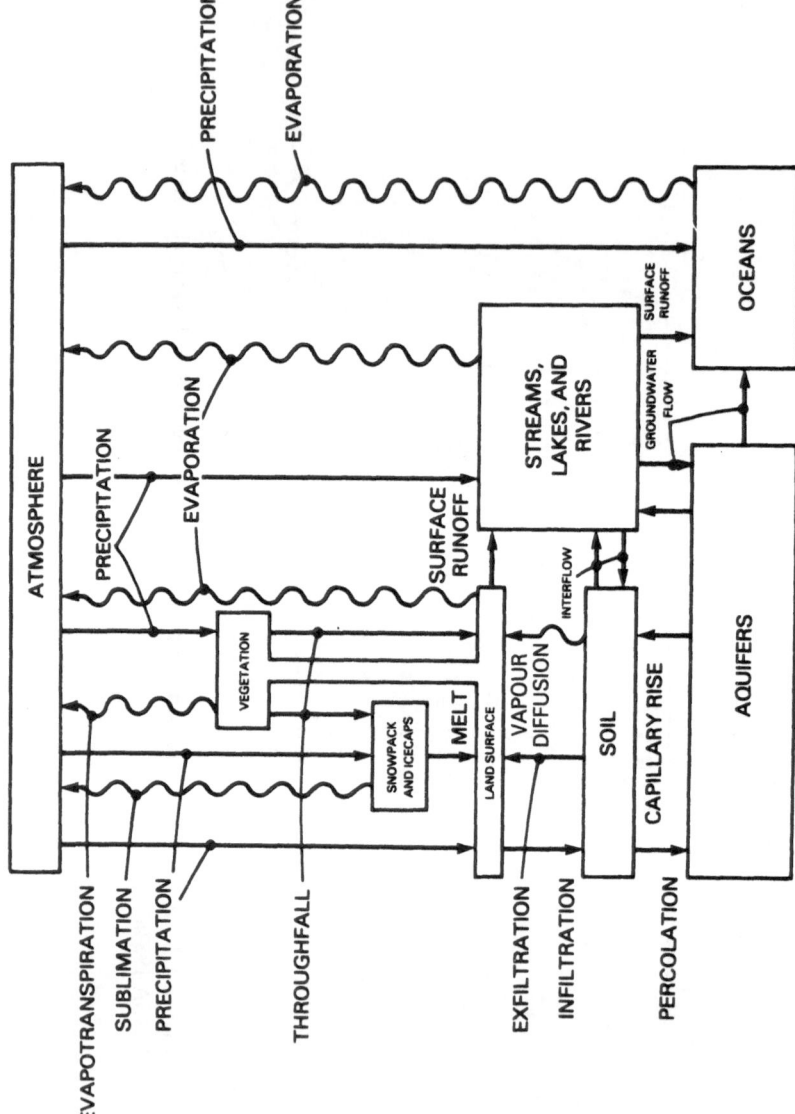

Figure 1.2 An engineering view of the hydrologic cycle (after Eagleson 1970).

Table 1.1 Waters of the global hydrosphere (L'vovich 1979) and fluxes of the global hydrologic cycle (Todd 1970)

Volume of waters	Volume of water (10^3 km)	Total volume (%)
World ocean	1 370 323	93.9600
Groundwater (including water of the zone of	60 000	4.1200
active water exchange)	4 000	0.2700
Glaciers	24 000	1.6500
Lakes	280	0.0190
Soil moisture	85	0.0060
Vapours in the atmosphere	14	0.0010
River water	1.2	100.0000
Total	1 454 193.2	100.0000
Global fluxes:		
Annual evaporation		
from world ocean	350	0.026
from land areas	70	0.005
Total	420	0.031
Annual precipitation		
on world ocean	320	0.024
on land areas	100	0.007
Total	420	0.031
Annual runoff to oceans from rivers and icecaps	38.0	0.003
Groundwater outflow to oceans[2]	1.6	0.0001
Total	39.6	0.0031

complicated by realizing that the readily accessible and dynamic parts of the water in the Earth–atmosphere system are a relatively small part of the total. Of the fresh water on the Earth, about one part in five is in liquid form. Of the liquid portion of this water that is fresh and unfrozen, nearly 99% is groundwater, about 1% is stored in lakes, 0.2% is stored as soil moisture, 0.1% is in flowing rivers at any one time, and about 0.005% is stored in plants.

There have been several prominent studies of the Earth's water balance reported in the literature (UNESCO 1978; I'vovich 1971; Nace 1972). Numerical simulations of the atmosphere have also provided a framework within which the dynamics of the hydrologic cycle can be examined (Manabe 1982). Other studies have emphasized how various characteristics of the Earth's surface affect the hydrologic cycle. For example, Mintz (1984) has shown that the distribution of precipitation, land cover and dynamic characteristics of the atmospheric circulation are clearly sensitive to changes in soil moisture.

Basin hydrologic cycle

The hydrologic cycle on the basin scale is not a closed cycle like the global one. Rather, the basin cycle is one in which input and output balance over some timescale. The process driving changes in the hydrology of the land surface can best be expressed in terms of the energy balance and the water balance at the surface. Figure 1.3 is a schematic of the hydrologic cycle on the basin scale. This figure emphasizes the various processes and represents the tremendous aggregation of these processes, which are extremely variable in space and time. Hydrologically, it is convenient to separate the processes into those that dominate during rain events (precipitation inputs, infiltration, runoff production and streamflow) and those that dominate during interstorm periods (evaporation, soil moisture redistribution and base flow). The dynamics of the hydrologic processes are influenced by the spatial and temporal characteristics of the inputs and by the controls exerted by the land surface. These controls concern elevation, soils, vegetation cover, man-made cover and the underlying geology. It is these processes that by and large have been the focus of hydrologists.

There are perhaps two reasons why hydrologists have concentrated on the basin-scale processes and controls. The first deals with the framework of problem solving. That is, the engineering problems of water supply, flooding, etc. have almost always been defined on a basin scale. The second reason deals with hydrologic instrumentation and its limitations. For the most part, hydrologic measurements (i.e. rain gauges, weirs, etc.) are point measurements that, in effect, sample a spatial phenomenon. As a result, their usefulness is limited to some undefined area around the point of measurement. Hydrologists have recognized these generally undefined limitations and have implicitly assumed that, at best, data collected at a point may be valid within that catchment but not in the neighbouring basin. This implicit feeling has been reinforced by the general lack of transferability of hydrologic data from one basin to another.

Thus the engineering approaches to hydrology have developed from a rather restrictive basin framework, although in all fairness it has generally been a very useful and successful approach. It was only very recently that this historic approach was shown to be limiting. With the 'shrinking' of our globe that has resulted from the improvement of communications, the burgeoning of population and the massive transfer of natural resources across so-called natural boundaries, people have rapidly become aware that hydrology, its problems and its benefits are not restricted to isolated basins. In addition, the developing awareness that other biogeochemical cycles are similarly global in nature has made it mandatory that scientists begin to look at the total globe. For example, the carbon cycle has recently been at the forefront of scientific and political discussions, driven primarily by the potential 'greenhouse effect'. A schematic of the carbon cycle is shown in Figure 1.4. It is important to note

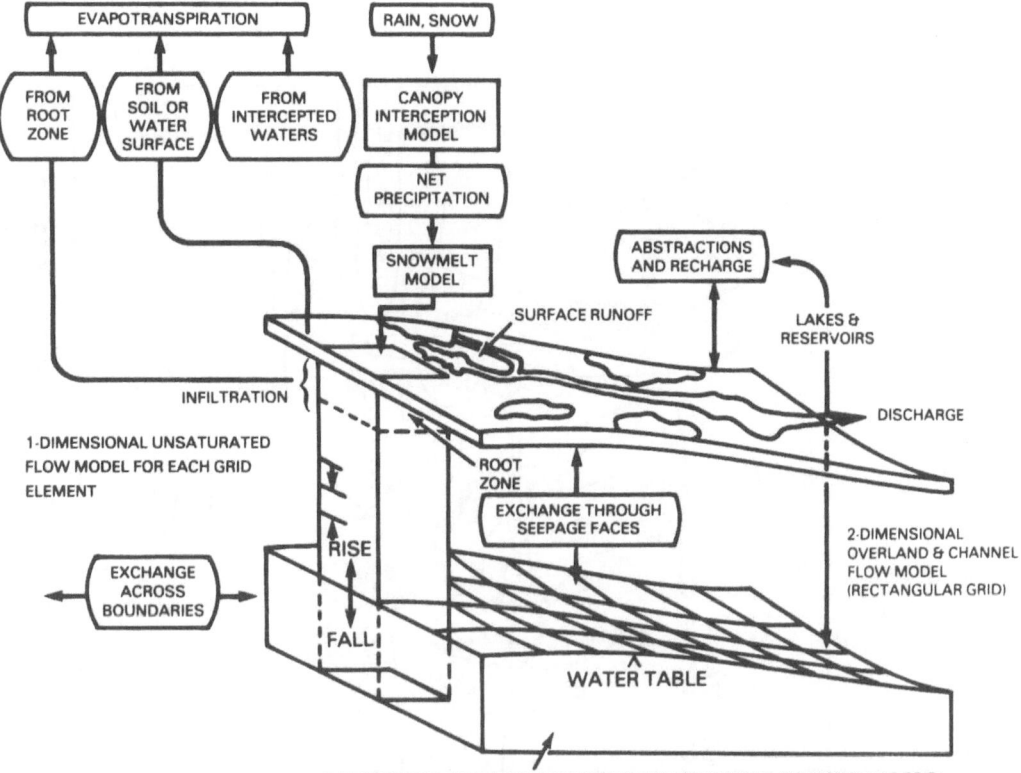

Figure 1.3 A schematic showing how distributed models represent the interactions of numerous hydrologic processes on a horizontal grid (after NASA 1987).

the dependence of the cycle upon water in several ways: as a transporting mechanism (involving eroded materials being carried as suspended sediment), as a reservoir, and in the photosynthesis process. The fact that the carbon cycle (and many other cycles) cannot really be separated from the hydrologic cycle truly emphasizes the need to understand hydrology on a global and local basis.

1.3   REMOTE SENSING IN HYDROLOGY

The rather recent development of Earth-oriented remote sensing instruments has enabled scientists, for the first time, to begin to study the global hydrologic cycle in a quantitative way. It has also enabled us to begin to study some of the previously insoluble problems of spatial variability. All of the benefits of remote sensing are not to be found some time in the future, however. Remote sensing is currently being used as an important source of data and information for hydrologic modelling and other water resource problems.

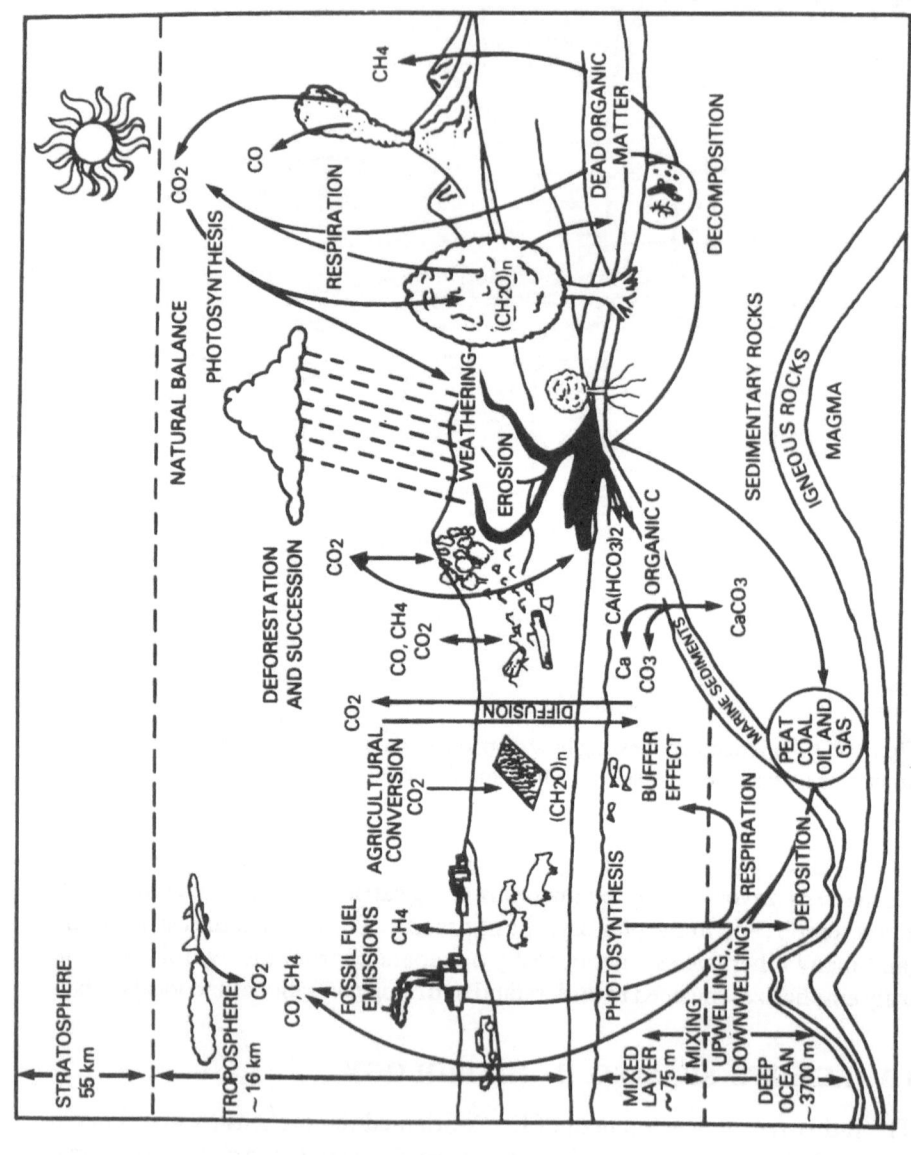

Figure 1.4 A schematic illustrating the carbon cycle and its dependency on the hydrologic cycle (after NASA 1987).

The potential for remote sensing and its application to hydrology is considerably greater than research has addressed so far. Measuring the characteristics of an area rather than a point, integrating several characteristics with one composite measurement, and improving prediction models with continuous or frequent feedback from satellite measurements are just a few of the aspects that must be explored further. Considerable research is needed if we are to fully realize the potential of remote sensing in hydrology. Treating remotely sensed data as a unique measurement of hydrologic characteristics offers the best opportunity for major advances in understanding the basin and global hydrologic balances.

The following chapters will address the state-of-the-art applications of remote sensing to the solution of current hydrologic problems. These chapters will also review current research so that, hopefully, readers can anticipate the new type of information that will some day be readily available.

REFERENCES

Dooge, J. C. I. (1973) *The nature and components of the hydrological cycle. Man's influence on the hydrological cycle, irrigation and drainage.* Paper No. 17, FAO, Rome, pp. 1–18.

Eagleson, P. S. (1970) *Dynamic Hydrology*, McGraw-Hill, New York.

L'vovich, M. I. (1971) The water balance of the continents of the world and the method of studying it. *Int. Assoc. Scientific Hydrology, General Assembly, Moscow.*

L'vovich, M. I. (1979) *World Water Resources and Their Future*, transl. American Geophysical Union, LithoCrafters, Chelsea, MI.

Manabe, S. (1982) Simulation of climate by general circulation models with hydrologic cycles. *Land Surface Processes in Atmospheric General Circulation Models* (ed. P. S. Eagleson), Cambridge University Press, New York, pp. 19–66.

Mintz, Y. (1984) The sensitivity of numerically simulated climate to land surface conditions. *The Global Climate* (ed. J. T. Houghton), Cambridge University Press, New York, pp. 79–105.

Nace, R. L. (1972) The present state and future prospects of global hydrology. *Proc. Symp. on the World Water Balance, Reading, England, July 1970*, UNESCO WMO.

NASA (1984a) *Earth Observing System, Working Group Report*, vol. I, NASA/Goddard Space Flight Center, Greenbelt, MD.

NASA (1987) *From Pattern to Process: The Strategy of the Earth Observing System*, vol. II, NASA.

Todd, D. K. (1970) *The Water Encyclopedia*, Water Information Center, Port Washington, NY.

UNESCO (1978) World water balance and the water resources of the earth. *Studies and Reports in Hydrology*, No. 25.

Vernadskii, I. (1960) *Izbrannye Sochineniya* (selected works), vol. 4, book 2, Izv. Akad. Nauk SSR, Moscow.

2

Basic principles of
remote sensing

2.1 INTRODUCTION

Remote sensing involves measurements of the electromagnetic spectrum that
can be used to characterize the landscape or infer properties of it (see Barrett
and Curtis (1982) for a complete description of environmental remote sensing).
Photography in the visible wavelengths was one of the first remote sensing
techniques to be used. Over the years remote sensing techniques have
expanded to provide the capability of making measurements over the entire
electromagnetic spectrum (Colwell 1983). Different sensors can provide
unique information about the properties of the surface or shallow layers of the
Earth. For example, measurements of the reflected solar radiation give
information on albedo, thermal sensors measure surface temperature, and
microwave sensors measure the dielectric properties of surface soil or snow.
The challenge for the remote sensing specialist and water resources scientist is
to interpret these remotely sensed properties in a way that can be used for
effective management and monitoring.

2.2 BASIC PRINCIPLES

The electromagnetic spectrum, which is shown in Figure 2.1, is the basis for all
remote sensing. Remote sensing takes advantage of the unique interaction of
radiation from the specific regions in the spectrum and the Earth.

There are four basic components of a radiation-based remote sensing
system:

1. radiation source (i.e. the Sun, radar);
2. transmission path (i.e. atmosphere, vegetation canopy);

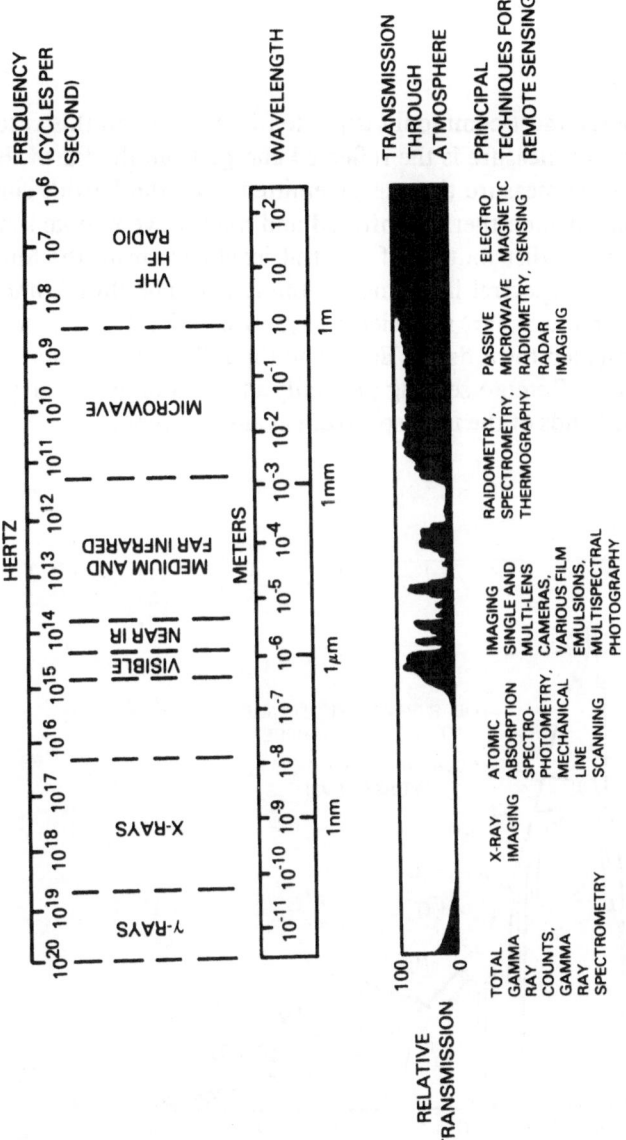

Figure 2.1 An illustration of the electromagnetic spectrum showing the relationship between wavelength and frequency, the common names for spectral bands, and the relative atmospheric transmission. The principal techniques used in remote sensing are also shown for their respective spectral region (after Barrett and Curtis 1976).

3. target (i.e. river, soil);
4. sensor (i.e. multispectral scanner, photographic film).

Each of these plays a significant role in either limiting or controlling what we can measure about the Earth's surface.

Radiation source

The radiation source most commonly exploited is the Sun. In this case, the characteristic that we measure is the reflected energy from the Earth, but in other applications we measure the energy emitted from the Earth's surface. These applications include thermal infrared and microwave remote sensing. Figure 2.2. shows the distribution of spectral irradiance from the Sun and Figure 2.3 shows the spectral irradiance of the Earth. The third commonly used radiation source is radar, in which energy from a limited region of the spectrum is propagated towards the Earth and the reflected or backscattered energy is measured. Remote sensing generally involves only one or several narrow regions or bands of the total spectrum at any one time.

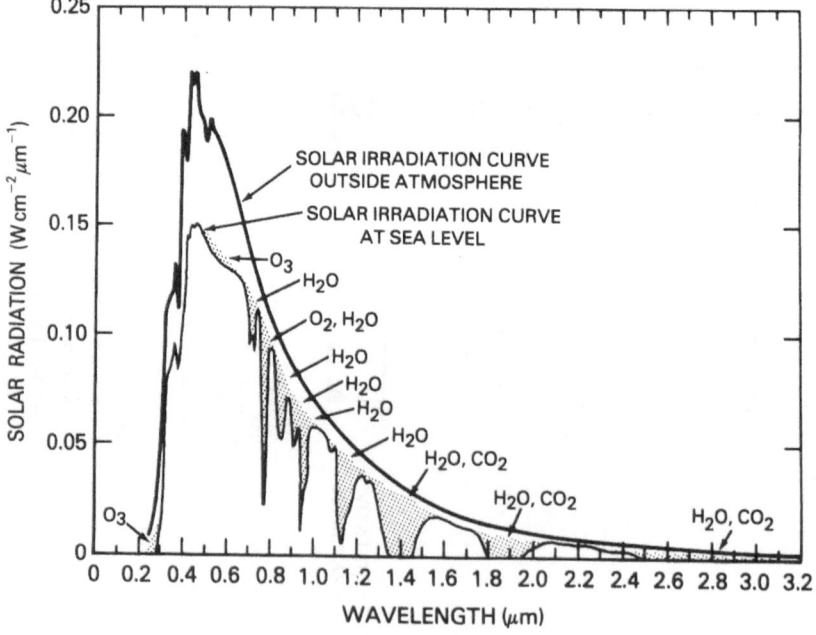

Figure 2.2 An illustration of the spectral distribution of solar radiation outside the Earth's atmosphere and what would be measured at sea level showing the effects of atmospheric absorption for water vapour, oxygen and carbon dioxide.

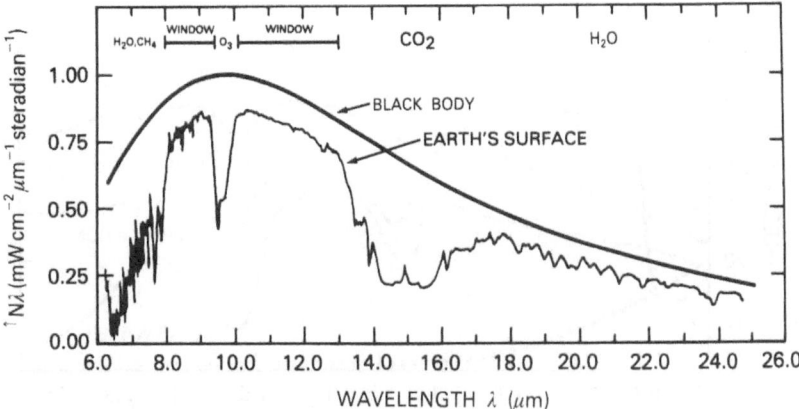

Figure 2.3 An illustration of the Earth's infrared radiance that would be measured outside the atmosphere compared with black-body radiance at 300 K.

Transmission path

Most often, the transmission path of the electromagnetic spectrum involves the atmosphere and this has a significant impact on which parts of the spectrum can be used. Specific gases in the atmosphere selectively affect the amount of energy that is transmitted, and this leads to a concept known as an *atmospheric window*. An atmospheric window is a wavelength band in which the atmosphere has little or no effect on the intensity of the Sun's radiation or reflected radiation from the Earth. Particulate matter such as smoke or dust can also affect the transmission path by scattering or absorbing radiation over the entire spectrum.

Target

The target is the subject of any observation as well as any other objects within the field of view of the sensor. Modern remote sensing is based on interpreting measurable variations in *spectral*, *temporal* and *spatial* characteristics of the Earth.

(a) Spectral characteristics (or signature) of the target are the unique spectral reflectances for specific Earth features. Figure 2.4 shows typical spectral reflectance curves for three basic types of Earth features: healthy green vegetation, dry bare soil (grey–brown loam) and clear lake water. The lines of this figure represent average reflectance curves, compiled by measuring a large sample of features. Note how distinctive the curves are for each feature. In general, the configuration of these curves is an indicator of the type and

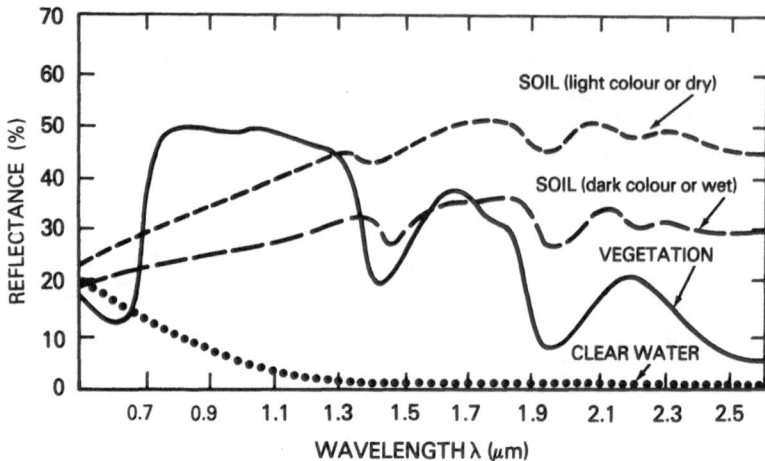

Figure 2.4   An illustration of the relative reflectance of typical ground cover types as a function of wavelength.

condition of the features to which they apply. Although the reflectance of individual features will vary considerably above and below the average, these curves demonstrate some fundamental points concerning spectral reflectance.

For example, spectral reflectance curves for healthy green vegetation almost always manifest the 'peak-and-valley' configuration illustrated in Figure 2.4. The valleys in the visible portion of the spectrum are dictated by the pigments in plant leaves. Chlorophyll, for example, strongly absorbs energy in the wavelength bands centred at about 0.45 and 0.65 μm. Hence, our eyes perceive healthy vegetation as green in colour because of the very high absorption of blue and red energy by plant leaves and the very high reflection of green energy.

The soil curve in Figure 2.4 shows considerably less peak-and-valley variations in reflectance. That is, the factors that influence soil reflectance act over less specific spectral bands. Factors affecting soil reflectance and moisture content include soil texture (proportion of sand, silt and clay), surface roughness, the presence of iron oxide, and organic matter content. These factors are complex, variable and interrelated.

Considering the spectral reflectance of water, probably the most distinctive characteristic is the energy absorption at reflected infrared wavelengths. In short, water absorbs energy in these wavelengths whether we are discussing water features *per se* (such as lakes and streams) or water contained in vegetation or soil. Locating and delineating water bodies with remote sensing data is done most easily in reflected infrared wavelengths because of this absorption property.

(b) *Temporal characteristics.*   Temporal effects are any factors that change the spectral characteristics of a feature over time. For example, the spectral characteristics of many species of vegetation are in a nearly continual state of change throughout a growing season. These changes often influence when we might collect sensor data for a particular application. Another example is the use of day–night or seasonal thermal data to infer information about the target. Temporal effects influence virtually all remote sensing operations. These effects normally complicate the issue of analysing spectral reflectance properties for Earth resources. Again, however, temporal effects might be the keys for gleaning the information sought in an analysis. For example, the detection of the process of change is predicted on the ability to measure temporal effects.

(c) *Spatial characteristics* that we use involve shapes and relative sizes as well as absolute sizes of objects. Spatial effects also refer to factors that cause the same types of features (e.g. corn plants) at a given point in time to have different characteristics at different geographic locations. An example of this type of spatial effect is the change in the leaf morphology of trees when they are subjected to some form of stress.

2.3 SENSOR

The type of sensor is perhaps the only characteristic of remote sensing over which the user has some control. Careful matching of the sensor to the problem can ensure that the results of the study will be useful and easily quantifiable. The commonly used sensors are briefly described below.

(a) *Gamma radiation.*   Gamma radiation remote sensing is based on the attenuation of natural terrestrial gamma radiation by soil water or a layer of snow (Carroll 1981). Many soils have a gamma radiation flux that originates from the naturally occurring ^{40}K, ^{238}U and ^{208}Ti radioisotopes. In a typical soil, 90% of the natural radiation originates in the top 20 cm of the soil (Zotinov 1968). The general procedure is to determine a background measurement for no snow or dry snow conditions. A subsequent measurement in the presence of snow or increased soil moisture will reveal an attenuated flux which can be related to the snow water equivalent or change in soil moisture.

(b) *Aerial photography.*   This has a long history of use in environmental management (Slama 1980). The first aerial photographs were of the visible portion of the electromagnetic spectrum. Advances in photography and films have led to the capability of making images of other parts of the spectrum with much interest in the near-infrared and thermal regions. For example, photographic images have provided information on sediment plumes, erosion features, discharges from pipes, and spills. Close-range stereophotographs have been used to study erosion and gully formation (Welch *et al.* 1984). Techniques for the interpretation of aerial photography have been well developed (Smith 1968; Slama 1980).

(c) Multispectral scanners. These are instruments that measure the spectral reflectance of narrow wavelength bands and record the information electronically. This technique involves measuring simultaneously the spectral response of the landscape in two or more narrow wavelength bands of the electromagnetic spectrum. Multispectral classification of these data is then used to discriminate objects based on their characteristic reflectances. Multispectral analysis has developed from early system, using two or more cameras with different lens filters to make images. Since 1972, the Landsat (originally ERTS) satellites have been providing four spectral bands of the Earth's surface with the multispectral scanner (MSS), and more recently with the thematic mapper (TM) (seven spectral bands). Slater (1985) has described 56 different multispectral imaging systems that have been used in the last decade for Earth resources studies. Multispectral scanner techniques have two major advantages over photographic methods. First, instruments can be designed to measure very narrow wavelength bands, and second, the digital form of the data enables the use of rapid and sophisticated analysis and classification techniques. Multispectral data have been analysed in many ways to produce information with potential use for the natural resource modeller. Examples of the use of multispectral data are to estimate land cover, land use, vegetation biomass, soil type, vegetation type, snow cover, water area, impervious area and various water quality parameters.

(d) Thermal sensors. Thermal remote sensors directly measure the emitted thermal energy of the Earth's surface. Surface temperature changes are the result of the balance of radiant, latent, sensible and ground heat fluxes (Price 1980). Analyses of remotely sensed thermal data (Price 1981) can be used to develop maps of the environmental conditions of the Earth's surface. In general, thermal sensors are used to measure variations in temperature across the landscape. One infers information about the properties of the landscape that affect temperature change. Examples of the use of thermal data are to estimate evapotranspiration, soil moisture, drainage patterns, groundwater seepage zones, canopy temperatures, and thermal plumes from thermoelectric power plants or industrial sources.

(e) Microwave sensors. Microwave remote sensing can directly measure the dielectric properties of the Earth's surface. Any change in these properties directly affects the reflectivity or emissivity measured by microwave systems. The dielectric property of the Earth's surface layer is in turn strongly

dependent on the moisture content. Thus, measurements in the microwave region of the electromagnetic spectrum can be related to the moisture content of the soil surface layers (Schmugge *et al.* 1980). Similar relationships exist for snow. The physical relationships between moisture, dielectric properties and microwave response, together with the ability of microwave sensors to penetrate cloud cover, make microwave sensors a useful all-weather sensor to measure the moisture of the Earth's surface. Meanwhile, active microwave systems (radar) send out an energy pulse and measure the Earth's naturally emitted microwave radiation. Active microwave and passive microwave systems have been flown on aircraft and satellites. Examples of the use of microwave data are to estimate soil moisture, vegetation type, snow water equivalent, condition of snowpack, frozen soil and sea ice.

(f) Lasers.   Laser remote sensing is a rapidly expanding research area with potential application to natural resource models. The principle involves projecting a narrow beam of coherent visible or near-infrared light and measuring the reflected energy with a photomultiplier tube to determine the distance between the laser system and the object of interest or to measure the backscatter from aerosols or the land surface. Airborne laser systems have been used to collect data for topographic maps and especially data for topographic surveys where high-density topographic information is required. Airborne lasers have also been used to estimate canopy height (Arp *et al.* 1982), terrain topography (Krabill *et al.* 1984), stream valley cross-sections (Krabill *et al.* 1984), air pollution (Sharp 1982) and chlorophyll (Hoge and Swift 1981). Laser systems may be very useful for environmental studies where topographic details, not available in published maps, are required. Examples of these needs would be erosion and gully studies, defining channel geometry, and topographic changes that may delineate erosion sources and sinks.

2.4 PLATFORMS

Platforms for supporting sensors can vary from ground-based supports to aircraft and satellites. Generally, truck-mounted ground-based and aircraft systems are used in sensor development to verify design characteristics and to learn how the sensor responds to the target characteristics. Figures 2.5 and 2.6 illustrate truck and aircraft experimental platforms. Truck and aircraft systems are too limited in their potential coverage and too expensive to be considered for operational remote sensing except for very limited and special applications. In almost all applications, the goal is eventually to get the sensor mounted on free-flying satellites. (Plate 1 shows a typical remote sensing satellite.) When this has been accomplished, the potential exists for obtaining large quantities of data over large areas of the globe, for extended periods of time at relatively low cost.

Figure 2.5 An illustration of a truck platform with passive microwave sensors on an experiment to measure soil moisture, also showing some of the micrometeorological equipment used (P. E. O'Neill, GSFC).

Orbital tracks are a variable that the user of remote sensing data must be aware of, although for the most part there is little that can be done about them. Satellite orbits are generally high-altitude geostationary, or low-altitude sun-synchronous with various repeat frequencies. A more complete description of satellites and their orbits can be found in Barrett and Curtis (1982) and Colwell (1983). Table 2.1 lists operational and planned satellites that have been, or could be, useful for hydrologic applications.

2.5 DATA ANALYSIS

Once data have been collected from a remote sensing system, the user must then interpret the data for his specific application. Generally, the data are available in three forms: as an image analogous to an aerial photograph, in an analogue format, and in digital format typically as a CCT (computer-compatible tape). These three data forms are interchangeable as imagery. Imagery can be digitized and digital data can be processed as an image or a

(a)

(b)

Figure 2.6 An illustration of the NASA DC-8 with (a) the airborne multifrequency, multipolarization radar for land studies and (b) the radar antenna in close-up (D. Evans, JPL).

Table 2.1 Planned US and foreign operational and research satellites for observing the earth

NOAA weather satellites—1978–1990s
  Objectives: operational weather data
  Orbit: sun-synchronous, 833–870 km; 7:00 a.m. and 2:00 p.m. equator crossing times
  Payload:

NOAA	8	9	G	10	I	11	K	L	I'	J'	K'	L'
Advanced very-high resolution radar (AVHRR)	×	×	×	×	×	×	×	×	×	×	×	×
High-resolution IR sounder (HIRS)	×	×	×	×	×	×	×	×	×	×	×	×
Stratospheric sounding unit (SSU)	×	×	×	×	×			×	×	×		
Microwave sounding unit (MSU)	×	×	×	×	×	×		×	×	×		
Data collection system (DCS)	×	×	×	×	×	×	×	×	×	×	×	×
Space environment monitor (SEM)	×	×	×	×	×		×	×	×	×	×	×
Solar backscatter UV exp. (SBUV)	×	×		×		×	×	×	×	×	×	×
Earth radiation budget (ERBE)	×	×	×	×					×	×		
Search and rescue (SAR)	×	×	×	×	×	×	×	×	×	×	×	×
Advanced microwave sounder (AMSU)							×	×	×	×	×	×
Data collection system	×	×		×			×	×	×	×	×	×
Advanced coastal zone colour scanner (ACZCS)								?		?		?
Planned or actual launch year	83	84	85	86	87	88	89	90	91	92	93	95
Equator crossing time	a.m.	p.m.	a.m.	p.m.	a.m.	p.m.	a.m.	a.m.	p.m.	a.m.	p.m.	a.m.

Instrument description:

AVHRR: 5 bands, 0.58–12.5 μm, 1 km/4 km resolution, 1600 km swath, temperature of clouds, sea surface and land

HIRS/2: 20 bands, atm. sounding, temperature and moisture profiles

SSU: 3 bands, atm. sounding, temperature profiles

MSU: 4 bands, 50.3–57.9 GHz, atm. sounding

DCS: random access from buoys, balloons and platforms

SEM: 3 instruments, solar protons, alpha particles and 'e' flux density

SBUV: 12 bands, 2550–3400 Å, solar spectrum, O_3 profiles, earth radiance spectrum

ERBE: determine earth's radiation loss and gain

AMSU: 20 bands, 10–90 GHz, possibly 180 GHz, all-weather temperature profiles

ACZCS: 9 bands, 0.4–0.88 μm, 10.5–12.5 μm, ocean colour, sea surface temperature

Argos data collection and position location system (Argos): 3 bands — 137.77, 136.77 and 401 MHz — platform location and sensor data relay

GOES — Geosynchronous weather satellite system — 1975–1990s

Objectives: operational weather data, cloud cover, temperature profiles, real-time storm monitoring, severe storm warning

Orbit: geostationary at east and west

Payload: visible and infrared spin scan radiometer (VISSR), VISSR atmospheric sounder (VAS), data collection system (DCS), space environment monitor (SEM)

1–5 inactive

6.6 failed

7I-M operational

Instrument description:

VISSR: 2 band, 0.55–0.70 μm, 10.5–12.6 μm, 0.9 km resolution visible, 8 km resolution IR, sensitivity of 0.4–1.4 K, day/night cloud cover, earth/cloud radiance temperature measurements

VAS: 12 bands, 0.55–0.70 μm, 3.9–14.7 μm, day/night cloud cover, atmospheric temperature and water content

DCS: random access from buoys, balloons and platforms

SEM: solar protons, alpha particles and 'e' flux density

DMSP — Defence meteorological satellite programme — 1970s–1990s

Objectives: operational weather data for DOD

Orbit: sun-synchronous, 720 km, equator crossing time: as desired

Payload: operational linescan system (OLS), multispectral IR radiometer (MIR), microwave temperature sounder (MTS), space environment sensor (SES) and special sensor microwave imager (SSM/I)

20 inactive; latest, 502-4, launched 3 February 1988

Instrument description:

OLS: 0.4–1.1 μm, 10–13 μm, 0.56 km/2.78 km resolution, global cloud cover

MIR: 9.8 and 12.0 μm bands, 13.4–15.0 μm (CO_2), 18.7–28.3 μm (H_2O), vertical temperature profiles

MTS: 50–60 GHz, band scanning microwave temperature sounder

SES: charged-particle monitor

SSM/I: 19.35, 37.0, 85.5 GHz, dual polarization, 22.23 GHz vertical polarization; 1400 km swath; precipitation, soil moisture, wind speed over ocean and sea ice morphology, water vapour

Landsat — 1972–1990s

Objectives: operational and data: vegetation, crop and land use inventory

Orbit: sun-synchronous, 705 km, 9:30 a.m. node, 16 day repeat

Payload: multispectral scanner (MSS), thematic mapper (TM)

1, 2 and 3 inactive

(*continued*)

Table 2.1 (*continued*)

4 and 5 operational
6 and 7 under development

Instrument description:
MSS: 5 band, 0.5–0.6 μm, 0.6–0.7 μm, 0.7–0.8 μm, 0.8–1.1 μm, 80 m resolution, 185 km swath
TM: 7 bands, 0.45–0.90 μm (4), 1.55–1.75 μm, 2.08–2.35 μm, 10.4–12.5 μm, 30 m/120 m resolution, 185 km swath

SPOT - Systeme probatoire, d'observation de la terre — 1986 launch with follow-on in 1991

Objectives: operational land use and inventory monitoring system
Orbit: sun-synchronous, 10:30 a.m. node, 2.5 day repeat
Payload: SPOT

Instrument description:
HRV (1,2): High-resolution visible imager (land surface topography and composition) 3 bands, 0.5–0.6 μm, 0.6–0.7 μm, 0.78–0.9 μm, 20 m resolution colour mode, 10 m resolution panchromatic mode (0.51–0.73 μm), 60 km × 60 km viewing area, swath of 950 km centred around nadir, stereoscopic images
DORIS (2,3,4,5): dual doppler receiver (precise orbit determination)
VMS (3,5): vegetation monitoring sensor
HRVIR (3,4,5): improved high-resolution visible imager

ERS-1 — ESA first remote sensing satellite — 1991 launch, ERS-2 in 1994, ERS-3 planned (European programme)

Objectives: coastal ocean and ice studies, global weather, land use
Orbit: sun-synchronous, 785 km, 10:30 a.m. equator crossing time, 3 day repeat cycle
Payload: active microwave instrument (AMI), along track scanning radiometer (ATSR)

Instrument description:
AMI: SAR: C-band 5.3 GHz, 30 × 30 m resolution, 80–200 km swath scatterometer (wind mode): 3 beam C-band, VV polarization, 500 km swath, 50 km resolution, range 4–24 m s^{-1}, accuracy 2 m s^{-1} or 10%; scatterometer (wave mode): 5 km × 5 km image every 100 km; altimeter: Ku-band (12.5 GHz), 10 cm precision — land, 40 cm precision — ocean, 1.2 m diameter antenna
ATSR: radiometer, 3.7, 11 and 12 μm bands, 1 km × 1 km resolution, 50 km swath;
PRARE: precision range and range rate experiment; laser retroreflector
LR: laser retroreflector (precise orbit determination)
RA (ERS-1): radar altimeter (ocean topography; ocean waves, ocean ice, ocean surface winds)

Radarsat — Canadian radar programme — 1994–2000, 3/4 satellite series

Objectives: high-resolution studies of Arctic area; agriculture, forestry and water resource management; ocean studies
Orbit: sun-synchronous, 792 km altitude, 3 day repeat cycle
Payload: synthetic aperture radar (SAR), optical sensor (TBD), microwave sensor (TBD)

Instrument description:
SAR: C- or L-band, 150 km swath, 25–30 m resolution, 4–100 km look angles

UARS — Upper atmospheric research satellite — 1991 launch
  Objectives: coordinated measurement of major upper atmospheric parameters
  Orbit: 57° inclination; 600 km altitude
  Payload: cryogenic limb array etalon spectrometer (CLAES), halogen occultation
    experiment (HALOE), high-resolution doppler imager (HRDI), improved
    stratospheric and mesospheric sounder (ISAMS), microwave limb sounder
    (MLS), particle environment monitor (PEM), solar stellar irradiance comparison
    experiment (SOLSTICE), solar-UV spectral irradiance monitor (SUSIM), wind
    measurement in the mesosphere (WINTER), active cavity radiometer irradiance
    monitor (ACRIM), solar backscatter UV experiment (SBUV)

  Instrument description:
  CLAES: global synoptic measurement of nitrogen and chlorine ozone-destructive
    species, minor constituents temperature
  HALOE: stratospheric species concentration
  HRDI: middle atmospheric winds
  ISAMS: atmospheric temperature and species concentration
  MLS: vertical profiles of O_3 and O_2, wind measurements, inferred pressure
  PEM: charged-particle entry measurements for atmosphere
  SOLSTICE: solar irradiance from 1150 to 4000 Å
  SUSIM: solar flux changes over 1250 to 4000 Å range
  WINTER: temperature, wind and OH concentration
  ACRIM: solar constant monitor
  SBUV: vertical O_3 distribution
  MEPS (medium-energy particular spectrometer): particles and fields environment,
    LeV–5 MeV
  AXIS (atmospheric X-ray imaging spectrometer): particles and fields environment;
    2–300 keV in 6 energy channels
  MAG (fluxgate magnometer): particles and fields environment
  HEPS (high-energy particle spectrometer): in 9 channels

TOPEX/POSEIDON — Topography experiment USA/France — 1992 launch

  Objectives: ocean topography, ocean current signatures
  Orbit: 1334 km, 63.13° inclination, 10 day repeat

Payload and instrument description:
  ALTIMETERS: 2 band, Ku 13.5 GHz, C 5.3 GHz, 2 cm precision, atmospheric
    correction provided by an on-board microwave radiometer
  LRA (laser retroflector array): precise orbit determination
  IMR (topex microwave radiometer): Atmospheric composition, earth radiation
    budget at 18, 21 and 37 GHz
  GPS (global positioning system): precise orbit determination

Meteor-2 — Operation weather satellite, USSR — 1977

  Objective: operational weather data
  Orbit: 900 km polar orbit, 2 day repeat, 2100–2600 km swath

(continued)

Table 2.1 (*continued*)

Payload: 3 channels: 0.5–0.7 μm, 2 km resolution, 0.5–0.7 μm, 1 km resolution, 8–12 μm, 8 km resolution

Meteor-priroda — Environmental satellite, USSR — 1991 launch

Objective: operational environmental, land use and inventory
Orbit: 650 km polar orbit
Payload:
MSU-M: 0.5–0.6 μm, 0.6–0.7 μm, 0.7–0.8 μm and 0.8–1.1 μm, all 1 km resolution, 1930 km swath
MSU-S: 0.5–0.7 μm and 0.7–1.0 μm, 240 m resolution and 1380 km swath
MSU-SA: 0.5–0.6 μm, 0.6–0.7 μm, 0.7–0.8 μm and 0.8–1.0 μm, all 170 m resolution, 600 km swath
Fragment: 0.4–0.8 μm, 0.5–0.6 μm, 0.6–0.7 μm, 0.7–1.1 μm, 1.2–1.3 μm, 1.5–1.8 μm and 2.1–2.4 μm, all 80 m resolution, 30 km swath
MSU-VA: 0.5–0.7 μm, 0.7–0.8 μm and 0.8–1.0 μm, all 30 m resolution, 30 km swath

Meteosat — Geosynchronous weather satellite — European Space Agency, 1978 with follow-on

Objective: operational weather data
Orbit: geostationary, 0°
Payload: 3 band, 0.4–1.1 μm, 2.5 km resolution, 5.7–7.1 μm, 1 km resolution, 10.5–12.5 μm, 5 km resolution

GMS — Geosynchronous meteorological satellite — Japan, 1977 with follow-on

Objective: operational weather data
Orbit: geostationary, 140° E
Payload: visible and infrared spin scan radiometer (VISSR), 2 bands, 0.5–0.75 μm, 1.25 km resolution, 10.5–12.5 μm, 5 km resolution

INSAT — Indian Weather Satellite — India, planned launch

Objective: operational weather data
Orbit: geostationary
Payload: very-high-resolution radiometer (VHRR), 0.55–0.90 μm, 2.75 km resolution, 10.5–12.5 μm, 11 km resolution

IRS — Indian remote sensing mission — India, not yet launched successfully

Objective: earth resources
Orbit: 904 km sun-synchronous, 22 day repeat
Payload: linear imaging self-scan (LISS), 0.45–0.52 μm, 0.52–0.60 μm, 0.65–0.69 μm and 0.76–0.90 μm, all 73 m resolution, 148 km swath

J-ERS-1 — Earth resources satellite — Japan, 1992

Objective: earth resources
Orbit: 570 km sun-synchronous, 42 day repeat
Payload: SAR, 1.23 GHz, 25 m resolution, 75 km swath, visible and near-IR radiometer (VNIR), 0.45–0.52 μm, 0.52–0.60 μm, 0.63–0.69 μm, 0.76–0.95 μm, all 25 m resolution, 150 km swath

computer-produced map; analogue data can be developed from one or the other, or used to produce digital data or an image. Usually, photometric accuracy is not preserved in analogue data and in conversions between analogue and digital forms unless calibration data are also provided and great care is maintained to preserve accuracy.

The conversion from imagery or analogous data to digital data, and vice versa, is based on separating the measured values of reflectance into binary increments, usually based on byte wordlengths with 0 representing the darkest or lowest level of reflectance and 255 the highest (brightest) reflectance.

Once the data are in digital form there are many instruments and computer systems that can perform a myriad of useful analyses to help the user interpret the data. Imagery from specific spectral bands or combinations of bands can be produced. Computer systems can also perform different types of classification procedures and compute different statistics of the scene. In addition the data may be included in numerical hydrologic models. Scene preprocessing that may be necessary includes normalization for sun angle, correction for atmospheric conditions, and geo-referencing the scene to a chosen map scale and coordinate system.

For more details on these aspects of remote sensing, readers are referred to the general publication of Lillesand and Kiefer (1979) and those of Colwell (1983), Swain and Davis (1978) and Barrett and Curtis (1982).

REFERENCES

Arp, H., Griesbach, J. C. and Burns, J. P. (1982) Mapping in tropical forest: A new approach using laser ARP. *Photogram. Eng. Remote Sensing* **48**, 91–100.

Barrett, E. C. and Curtis, L. F. (1982) *Introduction to Environmental Remote Sensing*, Chapman and Hall, London.

Carroll, T. R. (1981) Airborne soil moisture measurements using natural terrestrial gamma radiation. *Soil Sci.* **132**, 358–66.

Colwell, R. N. (ed.) (1983) *Manual of Remote Sensing*, vols I and II, American Society of Photogrammetry, Falls Church, VA.

Hoge, F. E. and Swift, R. N. (1981) Airborne simultaneous spectroscopic detection of laser-induced water Raman backscatter and fluorescence from chlorophyll-a and other naturally occurring pigments. *Appl. Opt.* **20**, 3197–205.

Krabill, W. B., Collins, J. G., Link, L. E., Swift, R. N. and Butler, M. L. (1984) Airborne laser topographic mapping results. *Photogram. Eng. Remote Sensing* **50**, 685–94.

Lillesand, T. M. and Kiefer, R. W. (1979) *Remote Sensing and Image Interpretation*, Wiley, New York.

Price, J. C. (1980) The potential of remotely sensed infrared thermal data to infer surface soil moisture and evaporation. *Water Resour. Res.* **16**, 787–95.

Price, J. C. (1981) Use of remote sensed infrared data for inferring environmental conditions from surface characteristics and regional scale meteorology. *Proc. 1981 Int. Geoscience and Remote Sensing Symp.* IEEE, Washington, DC, pp. 1195–201.

Schmugge, T. J., Jackson, T. J. and McKim, H. L. (1980) Survey of methods for soil moisture determination. *Water Resour. Res.* **16**, 961–79.

Sharp, B. L. (1982) Laser remote sensing of atmospheric pollutants. *Chem. Br. 1982*, 342–8.

Slama, C. C. (ed.) (1980) *Manual of Photogrammetry*, American Society of Photogrammetry, Falls Church, VA.

Slater, P. N. (1985) Survey of multispectral imaging systems for earth observations. *Remote Sensing Environ.* **17**, 85–102.

Smith, Jr, J. T. (ed.) (1968) *Manual of Color Aerial Photography*, American Society of Photogrammetry, Falls Church, VA.

Swain, P. H. and Davis, S. M. (eds) (1978) *Remote Sensing: The Quantitative Approach*, McGraw-Hill, New York.

Welch, R., Jordan, T. R. and Thomas, A. W. (1984) A photogrammetric technique for measuring soil erosion. *J. Soil Water Conserv.* **39**, 191–4.

Zotinov, N. V. (1968) Investigations of a method of measuring snow storage by using gamma radiation of the earth. *Sov. Hydrol., Sel. Pap.* **3**, 254–66.

3

Precipitation

3.1 INTRODUCTION

Precipitation is the input flux of the hydrologic cycle, whether it is for land processes or over the oceans. Precipitation occurs in two phases: in the liquid phase as rainfall and in the solid phase as snow and frozen rain. As precipitation is the input to the hydrologic cycle, it is vitally important to quantify it accurately. Any errors in estimating precipitation are propagated, and in some cases magnified, by other hydrologic processes. Determining the spatial and temporal depth of precipitation input to the Earth is necessary for everyday management of water resources such as rivers and reservoirs, irrigation, weather forecasting and predicting snowmelt runoff. It is also an essential component of scientific investigations of the hydrologic cycle, the global water balance and large-scale global atmospheric modelling. Indirectly, it is an essential component of all or nearly all biogeochemical cycles because most of these are driven by the water fluxes in the hydrologic cycle.

Historically, the estimation of precipitation has been accomplished by relatively simple instrumentation that samples the rain or snow by capturing a volume over a continuous or fixed time interval. This instrumentation, commonly referred to as a rain gauge, provides a fairly accurate measure of point rates and depths of precipitation. The major shortcoming of this instrumentation is that the measurement is only at a point; it has been well documented that rainfall on the Earth's surface varies greatly in both time and space from the shortest and smallest scales upwards. Although there are a vast number of rain gauges worldwide, even the relatively dense network such as that in West Germany, which averages one rain gauge every 100 square kilometres (Schultz 1989), and that in the United States, which averages about one gauge every 250 square kilometres (Linsley and Franzini 1955), are not adequate to define the precipitation input for most needs. The result of this is that rainfall can be measured relatively accurately for small areas with a network of rain gauges, but this approach is not practical for large areas, remote land areas of the globe and for oceans.

Recognizing the practical limitations of rain gauges, hydrologists have increasingly turned to remote sensing as a possible means for quantifying the precipitation input to the globe. Because the fundamental approach to measuring rainfall and snow are different with respect to remote sensing, snow will be discussed separately in Chapter 4.

3.2   GENERAL APPROACH

Direct measurement of rainfall from satellites for operational purposes has not been generally feasible because the opacity of clouds prevents direct observation of the precipitation with visible, near-infrared and thermal infrared sensors. However, improved analysis of rainfall can be achieved using both satellite and conventional ground-based data. Satellite data are most useful in providing information on the spatial distribution of potential rain-producing clouds, and gauge data are most useful for accurate point measurements. Ground-based radar has also proved useful for locating regions of heavy rain and for estimating rainfall rates. Useful data can be derived from satellites used primarily for meteorological purposes, including polar orbiters such as NOAA-N and DMSP, and geostationary satellites such as GOES, GMS and Meteosat, but their visible and infrared images can only provide information about the cloud tops. However, since these satellites do provide frequent observations, even at night with the thermal sensors, the characteristics of potentially precipitating clouds and the rates of changes in cloud area and shape can be observed. From these observations, estimates of rainfall can be made which relate cloud characteristics to instantaneous rain rates and/or rain totals over time.

The general approach used in making quantitative estimates of rainfall varies greatly depending on the type of remote sensing instrument used. Visible and infrared techniques, microwave radiometry, spaceborne radar and ground-based radar are the principal remote sensing approaches currently being tested or used. Each of these will be addressed in some detail below.

(a) *Visible and infrared techniques.*   The availability of meteorological and Landsat satellite data has produced a number of techniques for inferring precipitation from the visible and/or infrared (VIS/IR) imagery of clouds (see Barrett and Martin 1981; D'Souza and Barrett 1988). These techniques have led to the development of three dominant approaches: a cloud indexing approach, the thresholding approach, and the life-history approach. Cloud indexing, which is time independent, identifies different types of rain clouds and estimates the rainfall from the number and duration of clouds or their area. Thresholding techniques consider that all clouds with low upper-surface temperatures are likely to be rain clouds. Life-history methods are time dependent and consider the rates of change in individual convective clouds or in clusters of convective clouds. All such methods are essentially empirical in that they use statistical coefficients based on historical cloud- and ground-measured rainfall.

(b) *Microwave radiometry.* Microwave techniques offer a great potential for measuring precipitation because at some microwave frequencies clouds are essentially transparent, and the measured microwave radiation is directly related to the raindrops themselves. Remote sensing in the microwave region is generally practised at frequencies that take advantage of the peaks and valleys of the atmospheric attenuation curve (Figure 3.1). Measurements near the peaks are used for monitoring the atmospheric conditions, that is water vapour and temperature. Measurements at frequencies near the valleys of the attenuation curve are chosen to minimize the atmospheric effects and monitor features on the Earth's surface such as snow cover and soil moisture, or, in this case, precipitation.

Microwave radiometry or passive microwave techniques react to the rain in two fundamental ways: by emission/absorption and by scattering. With the emission/absorption approach, rainfall is observed through the emission of thermal energy by the raindrops themselves. This technique performs best over a uniformly cold background, such as could only be provided by the oceans. At lower satellite-observed microwave frequencies, for example at 19.35 GHz, rain is highly absorptive and will produce a warmer brightness temperature compared with the cold background (Wilheit *et al.* 1977). With the scattering approach the rain attenuates upwelling radiation from the Earth's surface and scatters or reflects cold, cosmic background radiation to the radiometer antenna. The scattering approach is not as background dependent, but it is less direct because frozen drops aloft are the main scatterers, and these are not as directly related to the actual falling rain. At frequencies greater than 80 GHz, say, scattering dominates and the rain produces a cooler brightness temperature compared with a warm background. In both cases the emission/absorption and scattering increase with increasing rain rates. Considerable effort is now being spent on the development of passive microwave rainfall algorithms using dual-frequency or multifrequency principles, or polarizations at a single frequency. The results are promising (Spencer *et al.* 1988), although for operational applications in the near to mid-term future it appears as if the most successful operational methods for general

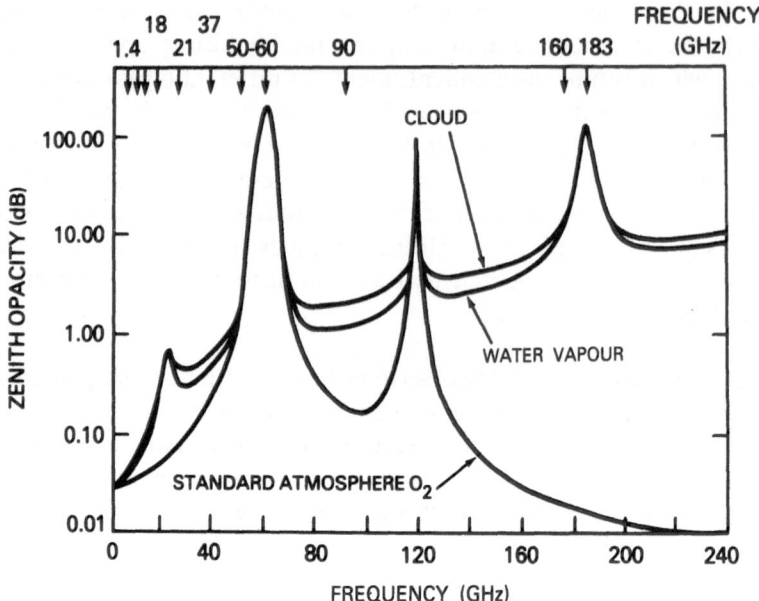

Figure 3.1 An illustration of microwave absorption in the atmosphere. The lower atmospheric opacity curve shows the effect of oxygen. The middle opacity curve shows the effect of adding 20 kg m^{-2} of water vapour to the oxygen. The upper opacity curve shows the effect of 0.2 kg m^{-2} stratus cloud added to the oxygen and water vapour (after NASA 1987).

rainfall measurement will likely use passive microwave data in conjunction with visible and infrared (Barrett *et al.* 1988).

(*c*) *Spaceborne radar.*   Radar measurement of rainfall is based on the Rayleigh scattering caused by the interaction of rain and the radar signal. Assuming the rainfall uniformly fills the radar pulse volume, the power of the radar pulse returned to the antenna can be expressed as:

$$Pr = C_1 C_2 \ K \ \Sigma \ D^6 / r^2 \tag{3.1}$$

where Pr is the average power returned from the precipitation at distance r, the radar target range, C_1 is an instrument parameter, C_2 is related to the dielectric constant of the precipitation (approximately a constant), K is an attenuation factor and D is the raindrop diameter. The radar response to rainfall is wavelength dependent according to D^6/λ, where λ is the wavelength. Because the radar interacts primarily with the rain, it can identify the presence of rain below the clouds and theoretically, at least, provide an opportunity to measure rainfall rates without several restrictive assumptions.

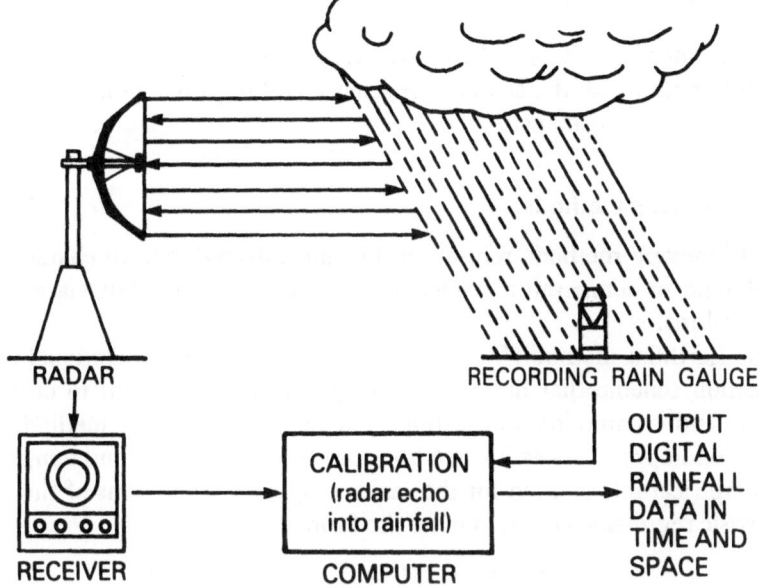

Figure 3.2   Schematic ground-based radar rainfall measurement (after Schultz 1988).

(*d*) *Ground-based radar.*   Ground-based radar is conceptually similar to spaceborne radar except that the radar is stationary and its area of measurement is limited to a circle with a radius up to about 100 km (Schultz 1989). Figure 3.2 illustrates the operation of a ground-based radar for measuring precipitation. A recording rain gauge is desirable for calibration. Thus, the rain rate, R, can be estimated according to

$$Z = aR^b \qquad (3.2)$$

where Z is the measured radar reflectivity and a and b are calibration parameters.

With ground-based radar there are two basic approaches (Dahlstrom 1985) used quantitatively to estimate precipitation.

1. Single-parameter measurement: this uses either the backscattered radiation which is related to precipitation intensity or the attenuation rate which is related to the precipitation rate.
2. Multiparameter measurement: here multiwavelength radar backscatter is related to the precipitation rate with the dual-polarization capabilities being used to give information on the drop size distribution.

3.3 CURRENT APPLICATIONS

From an increasingly wide range of available techniques, a few have been selected to illustrate the state of the science and to provide insight into their structure and performance.

Cloud indexing methods

Cloud indexing methods rely on visible and infrared data to characterize a cloud type or temperature which is then related to rainfall via empirical relationships.

The EarthSat method (Moses and Barrett 1986) is an operational rainfall estimation scheme that has been developed to provide input to crop yield models and commodity forecasting systems. The EarthSat method uses a regression approach to estimate 6 hour precipitation from cloud temperature and empirical information for the major crop-growing regions of the world. The basic regression equation takes the form

$$R = [\alpha + \beta(C \times V \times 0.6)]M \qquad (3.3)$$

where R is the 6-hour rainfall (mm), α and β are rainfall coefficients for a given region of the world (see Table 3.1), C is the local cloud brightness category determined from the infrared satellite imagery and, from Table 3.2, V is a vertical motion class determined from the imagery or from Table 3.3, and M is a weighting factor dependent on climate. The results of the regression calculation can be further modified by consideration of synoptic station reports. Experienced meteorologists can usually improve upon these estimates by improving the vertical motion fields through interpretation of the satellite imagery.

The manual Bristol method uses an empirical relationship between satellite-determined cloud indices, climatic indices dependent on the mean monthly

Table 3.1 Earth satellite-alone rainfall coefficients (Moses and Barrett 1986)

Region	Country identifier	Rainfall coefficients (α, β) Morning	Afternoon
USA	1	− 4.65, 1.88	− 12.43, 3.23
South America	2	− 4.90, 2.30	− 14.53, 2.61
Western USSR	3	− 2.65, 1.20	− 6.40, 2.03
Eastern USSR	4	− 2.30, 1.31	− 6.20, 2.32

Table 3.2 Image conversion chart (Moses and Barrett 1986)

Count value	Approx. temperature (°C)	Average rain (mm)	Brightness level
>218	<−56	25.0	6
210–217	−51−−55	16.0	5
199–209	−45−−50	9.0	4
154–198	−25−−44	4.0	3
125–153	−12−−24	2.0	2
112–125	−7−−11	0.0	1
>112	<−6	0.0	0

Table 3.3 Vertical motion classes (Moses and Barrett 1986)

Class	Vertical $(cm\,s^{-1})$
1	<−4
2	−3−−1
3	−1–1
4	1–3
5	>+4

rainfall, and 12-hour rainfall totals. A family of curves (Figure 3.3) has been developed from a number of studies in tropical and mid-latitude zones (Barrett 1981). These studies indicated a consistent increase in precipitation amounts from dry to hot–humid climates, but also indicated that higher-intensity rain clouds could not often be differentiated from lower-intensity rain clouds. Hence there is a need to treat each pixel location separately and in the light of climatological information. Table 3.4 lists the types of clouds and their associated indices.

The curves in Figure 3.3 can be used in two ways, depending on the availability of surface data. For the first case, when surface reports are available, these data are used to estimate precipitation amounts for surrounding areas with no surface records, that is 'areas of influence' around each gauge or set of gauges. Within each of these, the station rainfall is normalized by the mean monthly rainfall per rain day and a morphoclimatic weight. The appropriate curves in Figure 3.3 are then used with the cloud-type

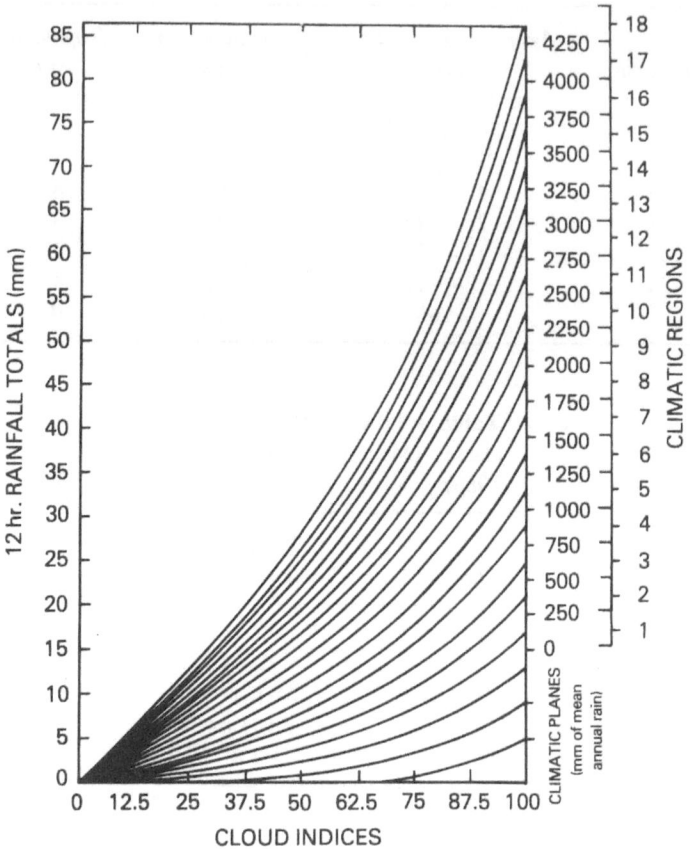

Figure 3.3   The Bristol 'global regression' diagram (after Moses and Barrett 1986).

index for that area and the normalized rainfall to adjust the surface records for rain cloud characteristics. In the second case, when there are no surface station reports, the satellite image is analysed manually to determine the cloud type and aerial coverage. The rainfall is then estimated on a pixel-by-pixel basis according to the cloud type and mean monthly rainfall from Figure 3.3.

The Bristol method has been further developed into an interactive method known as BIAS (Bristol/NOAA interactive system) originally for the USDA AgRISTARS programme (Barrett *et al.* 1986). BIAS uses the following relationship:

$$R = f(C_t, C_a, C_d, C_c, S_w) \qquad (3.4)$$

Table 3.4 Bristol cloud types (from Moses and Barrett 1986)

Name	Type	Common characteristics
Cumulonimbus with cirrus	10	Overshooting/colder tops (active area)
Cumulonimbus	8	Well-developed convective clouds with/ without anvils
Layered stratus with embedded cumulonimbus	7	Bright, banded frontal cloud with pebbled texture in visible
Layered stratus	5	Bright, banded frontal cloud
Cumulus congestus	3	Patches of moderately bright convective
Layered cumulus		clouds; disintegrating frontal cloud
Thick altostratus	2	Downwind borders of warm fronts and warm occlusions
Stratus	1	Thin low-level cloud sheets; decaying
Stratocumulus debris		remains of deep convective rains

where R is the 6-, 12- or 24-hour rainfall, C_t is a rain cloud type, C_a is the fraction of rain-cloud-type area, and C_d is the duration. The product of C_a, C_t and C_d is the 'cloud index' which is translated into rainfall estimates through the regression equations. Each pixel is assigned a climate category, C_c, in order to select the proper cloud index/rainfall regression equation to be used in the absence of surface measurements of rainfall. However, if the surface measurements are available, the synoptic weather term, S_w, may be used to provide more accurate results.

Thresholding methods

PERMIT (polar orbiter effective rainfall integrative technique) (Barrett *et al.* 1987) is based on temperature thresholding of thermal infrared (IR) imagery analysed by computer to identify potential rain clouds. The relationship used in PERMIT can be expressed as

$$R = f(C_T, C_d, M_c, S_w) \qquad (3.5)$$

where R is the estimated rainfall for periods of 10 days or longer, C_T is the cloud top temperature, C_d is the number of satellite observed rain cloud days, M_c is an empirical term that describes the combined climate/terrain effect and S_w is a synoptic weight related to surface measurements.

ADMIT (agricultural drought-monitoring integrative technique) (Barrett *et al.* 1987) is an extension of PERMIT that can be used to reduce the problem of thick cirrus clouds being misinterpreted as rain clouds. ADMIT uses daytime visible and IR image pairs to establish rain or no-rain thresholds. These are then used to reduce possible misinterpretation of rain clouds from the night-time IR imagery above. The ADMIT relationship can be written as

$$R = f(C_B, C_T, C_d, M_c, S_w) \qquad (3.6)$$

where C_B is the cloud brightness determined from the daytime visible imagery. All other terms are the same as in PERMIT.

Life-history methods

Some life-history methods are designed to provide rain estimates from any type of convectional clouds by taking into objective consideration the growth or dissipation of individual clouds with time. This approach implicitly recognizes that convective clouds exhibit different rainfall intensities during their growth and dissipation cycle. The ERL (Environmental Research Laboratory) or Woodley–Griffith technique was developed initially to predict rainfall over south Florida as part of the Florida area cumulus experiment (Griffith *et al.* 1978). The ERL method uses an empirically derived relationship between calibrated ground-based radar echoes and geostationary satellite imagery of cloud areas. A time-history relationship between the radar echo area and the cloud area is developed for discrete time intervals during the lifetime of the cloud. The relationship used in the ERL technique can be written as:

$$R_v = I \times A_e \times \Sigma\, q_i\, b_i \times \Delta t \qquad (3.7)$$

where R_v is the rain volume in millimetres, I is the rain rate in $m^3\ km^{-2}\ h^{-1}$, A_e is the inferred radar echo in square kilometres, i is a temperature index, q is the fraction of pixels are each threshold within the cloud, b is a temperature-dependent weighting factor and Δt is the time interval between successive satellite images. Values of the rain rate, I, are obtained from a table listing I as a function of increasing or decreasing radar echo. The rain volume computed for each cloud is then apportioned over a grid array of hourly accumulation.

A family of life-history techniques with the more specific purpose of evaluating and monitoring extreme (high-intensity) events has been developed from the work of Scofield and Oliver (1977). By originally using half-hourly rainfall amounts for convective systems from tropical air masses, an analyst can then use a decision tree to make rainfall estimates at different points. This technique is divided into three parts: first, the active portion of the convective system is delineated; second, an initial estimate of rain rate is made from the IR image alone; then, third, the changes in two consecutive images (visible and IR) are evaluated to find clues that would indicate heavier rainfall. Figure 3.4(a–f) illustrates the enhanced infrared imagery used to estimate the rainfall from a storm over central Mississippi in April 1979. The comparison of

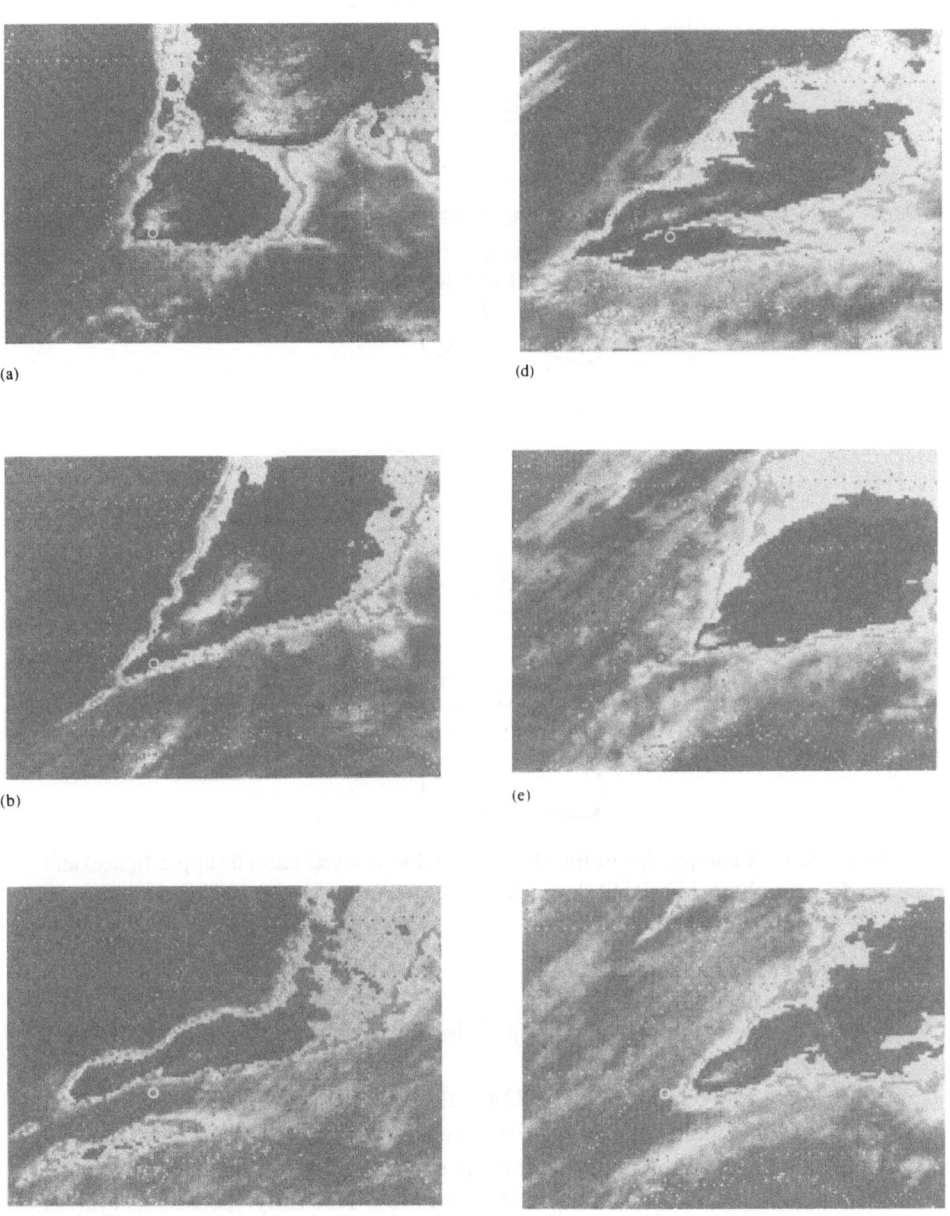

Figure 3.4 GOES data time series used to estimate rainfall shown in Figure 3.5 (Scofield and Oliver 1981). Enhanced infrared imagery: (a) 0430 GMT, 12 April 1979; (b) 0900 GMT, 12 April 1979; (c) 1500 GMT, 12 April 1979; (d) 1800 GMT, 12 April 1979; (e) 0100 GMT, 13 April 1979; (f) 0400 GMT, 13 April 1979.

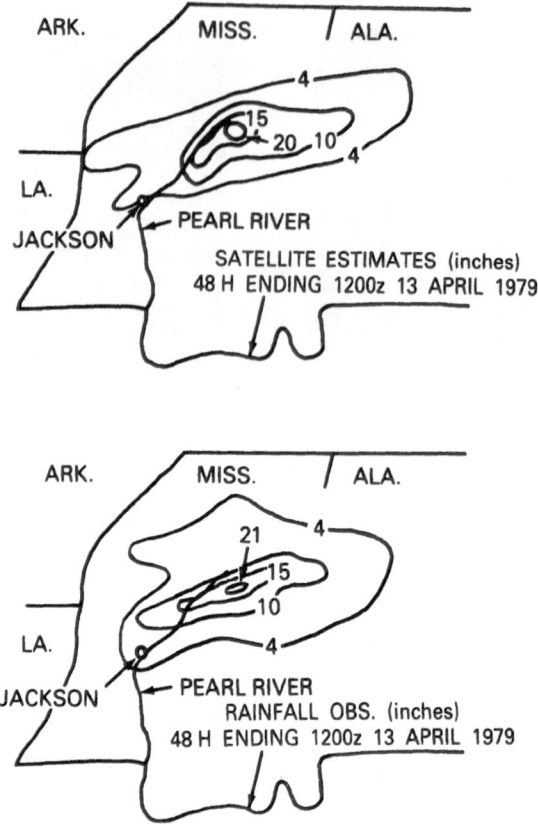

Figure 3.5 A comparison of the 48-hour satellite-derived rainfall (upper figure) and the 48-hour observed rainfall (lower figure) in inches ending at 1200 GMT, 13 April 1979 (after Scofield and Oliver 1981).

satellite-derived rainfall estimates and the observed rainfall is shown in Figure 3.5.

Based on this technique the NOAA flash flood programme in the United States now uses half-hourly GOES data to develop real-time estimates of heavy precipitation on an interactive computer system known as the interactive flash flood analyser (IFFA) (Clark and Morris 1986). This fully operational system has the capability to assimilate data from the GOES satellite together with ground-based and atmospheric data to develop precipitation estimates for input to a river forecast model.

Scofield (1986) has developed a series of seven convective and five extratropical cloud categories that can be used to help meteorologists improve their estimates of heavy precipitation across a range of different weather situations. The categories have been developed from the GOES data, from radar, surface and upper-air data, and from precipitation characteristics. Each category is based on the life history of the cloud pattern and cloud-top temperature changes as well as the other information.

Spaceborne microwave techniques

Microwave techniques have a great deal of promise for measuring rainfall because of the potential for sensing the rain itself and not a surrogate of rain such as the cloud type. Microwave radiation with wavelengths of the order of 1 mm to 5 cm results in a strong interaction between the raindrops and the radiation. This is because the drop size is comparable with the wavelength. Both passive systems (radiometers) and active systems (radar) have attributes that can be useful for measuring precipitation.

The approaches used in rainfall monitoring with passive microwave techniques are based on the frequency and/or polarization of the emitted radiation. From a practical point of view, at rain rates greater than a few millimetres per hour, scattering becomes significant and must be accounted for together with absorption. Wilheit *et al.* (1977) demonstrated the use of a radiative transfer model to estimate the higher rainfall rates. The model was verified using the ESMR-5 19.35 GHz data, a meteorological radar in Florida, USA, two rain gauges, and an upward-looking radiometer (also at 19.35 GHz). The results are shown in Figure 3.6. This technique has also been used to estimate weekly, monthly and annual rainfall maps for the major ocean areas.

The more difficult problem of measuring rainfall over land was documented by Meneely (1975), who showed that the ESMR-5 (19.35 GHz) measurement of upwelling radiation from land is only weakly affected by rain. However, Savage and Weinman (1975), using the 37.0 GHz channel from ESMR-6, showed that scattering by rain over land is strong enough to make at least qualitative estimates of rain coverage. Later, Weinman and Guettner (1977) showed that the upwelling radiation from rain should be unpolarized in contrast to the background radiation from the wet land surface, opening up a new range of possibilities for the development of rainfall algorithms.

A more recent rainfall algorithm is based on the difference in measured brightness temperatures at two frequencies, as suggested by Grody (1984). This algorithm is based on the relationship between emissivity and frequency, which decreases for most surfaces but increases for dry snow, old sea ice and in the presence of scattering caused by raindrops. Grody's algorithm can be written as

$$R = T_b(18) - T_b(37) \tag{3.8}$$

In equation (3.8), $T_b(18)$ and $T_b(37)$ are the measured horizontally polarized brightness temperatures at 18 and 37 GHz, respectively. When R is negative

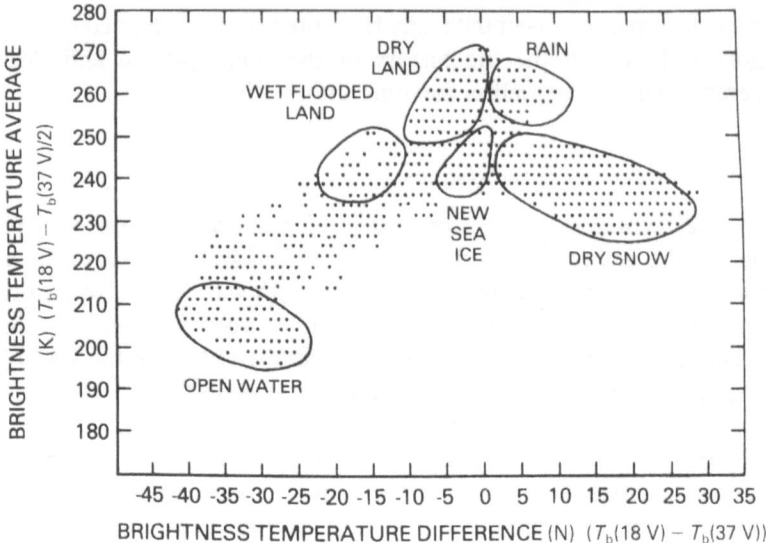

Figure 3.6 Plots of the 18 and 37 GHz vertical polarization brightness temperature average against difference for 20 January 1979 based upon SMMR and ground truth between 25 and 60 °N latitude. Six different geophysical parameters are identified based upon the clustering of the observations and comparison against surface data (after Ferraro *et al.* 1986).

the presence of rain is usually indicated. In later work, Ferraro *et al.* (1986) developed a classification approach for identifying rain as well as other geophysical features. Their classification scheme is based on the difference in two vertically polarized frequencies plotted against their average. Figure 3.7 illustrates this classification approach.

This frequency algorithm has been modified by Spencer *et al.* (1988) to include empirical coefficients in the following manner:

$$R = f(a \times T_b (18) - b \times T_b (37) + c) \tag{3.9}$$

where *a*, *b* and *c* are determined from regression analysis with surface measurements or ground-based radar.

The polarization algorithm suggested by Grody (1984) is based on the correlation between two polarizations at the same frequency so that the surface emissivity effects are minimized and the precipitation effects enhanced. This algorithm can be written as

$$\frac{T_v - BT_h}{1 - B} = T_u + vT_s \tag{3.10}$$

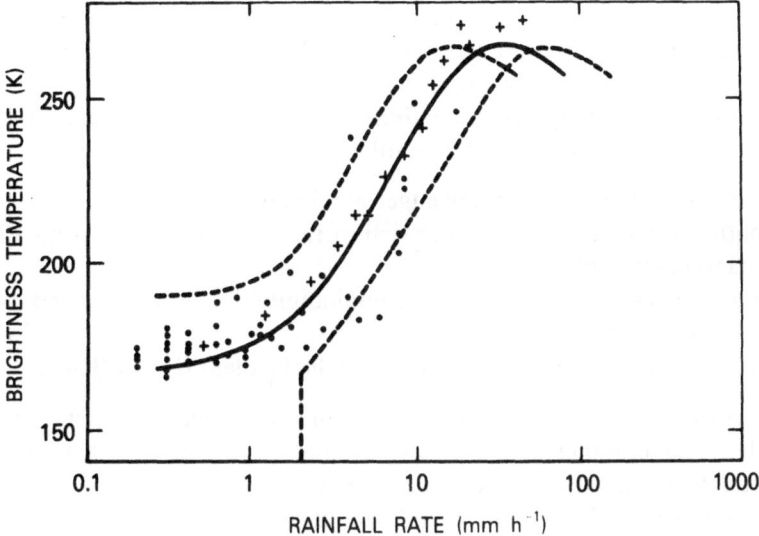

Figure 3.7 The change in brightness temperature at 19.35 GHz as a function of rainfall rate over the ocean (after Wilheit *et al.* 1977).

where T_v and T_h are the vertical and horizontally polarized brightness temperatures at a given frequency, v is the atmospheric transmittance, T_u is the upwelling radiation, T_s is the surface temperature and B is a frequency-dependent constant that is empirically determined for a given area and season.

Although the relationship between the rain rate and the microwave response is now fairly well understood (Atlas and Thiele 1981), there are several limitations to this approach. One of the more serious is the unknown depth of the liquid rain layer, especially in areas where warm rain forms without an ice phase. The effect of cloud density (made up of small droplets less than 50 μm in diameter) has not been accounted for and is a potential source of error that could perhaps be solved with multifrequency measurements. There is also work to be done on improving the modelling of the scattering mode to represent more realistic ice particle geometries. Finally, these problems do not begin to address the more fundamental problems of measuring rainfall. These include the great spatial and temporal distribution of rainfall and the fact that instantaneous rain rates are being measured when, in fact, what is needed is the integrated rainfall volume over some time period. Therefore, combining VIS, IR and passive microwave data is the next logical step in the development of rainfall algorithms. In this respect it is especially unfortunate that there are, as yet, no suitable passive microwave sensors on geostationary satellites.

In the first examination of the potential of combining VIS, IR and passive microwave (PMW), a hybrid approach (Barrett and Kidd 1987) investigated the use of SMMR (scanning microwave multichannel radiometer) passive microwave data in support of BIAS (discussed above) for rainfall monitoring in several areas of the globe where they could overlay the SMMR data on to the BIAS images. Although not a definitive study, it was concluded that the passive microwave data could be useful in helping to:

1. locate the leading and trailing edges of rain areas;
2. confirm the extent and organization of rain areas especially if no VIS/IR data were available;
3. locate heavy-rain areas where cumulonimbus cells are located within stratiform clouds;
4. locate areas of rain that could not normally be identified by BIAS.

Spaceborne radar has not yet been used for estimating precipitation for the simple reason that there is no existing satellite radar with suitable frequencies for rainfall. What has been produced is a fairly impressive body of information based on ground-based radar and the theory and planning of satellite programmes for measuring rainfall.

Meneghini *et al.* (1983) have proposed using a spaceborne radar for estimating rain rates based on the attenuation of the radar signal by the rain. They propose using the surface as a reference target for determining the path-averaged rain rate that is independent of the reflectivity/rain rate law and the radar calibration.

The potential for using dual-wavelength radar for measuring rain rates has been proposed by Goldhirsh (1988). He analysed the algorithms necessary to retrieve rain rate profiles from a spaceborne radar operating at 13.6 and 35 GHz by comparing the results with simulated rain rate profiles. It was shown that the backscatter method at 13.6 GHz is useful near the top part of the rain but that backscatter at 35 GHz is not useful nor is the attenuation approach. The dual-wavelength approach yielded acceptable accuracies for rain from about 0.75 to 3 km above the Earth. He concluded that 'no single technique gives rise to a panacea in the making of accurate rain measurements and that difficulties exist with each method'.

Ground-based microwave techniques

Although satellite techniques are still mainly experimental or developmental, ground-based radar has been used increasingly for operational rain and flood forecasting since early in World War II, as well as providing valuable research results for future ground and spaceborne systems. Successful applications have demonstrated a variety of approaches, from using calibrated radar systems alone, to supplementing the radar with satellite data, to using a range of single-frequency to multifrequency/multipolarization data.

Schultz (1988) has demonstrated the usefulness of radar data and their

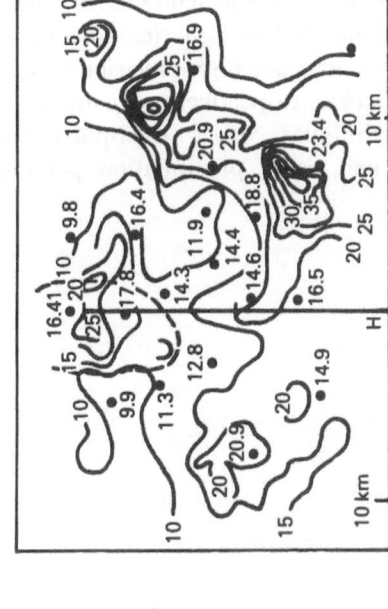

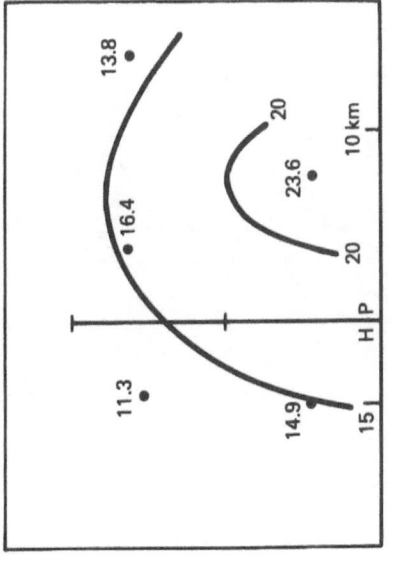

Figure 3.8 An illustration of area rainfall estimates based on point and radar measurements (28 July 1972, 20:20–0:20) (after Schultz 1988).

ability to provide better spatial definition of individual storm events. Figure
3.8 illustrates this comparison for a storm in West Germany in July 1972.

Goldhirsh *et al.* (1987) have shown how a dual-polarization radar can be used
to estimate rain rates. They compared the light (less than 7 mm h^{-1}) rain rate
measurements from the radar to low- and high-precision rain gauges to achieve
differences of about 5% and 16%, respectively. A comprehensive review of
differential reflectivity techniques using dual-polarization radar for
determining rainfall rates and other precipitation parameters has been
prepared by Ulbrich (1986).

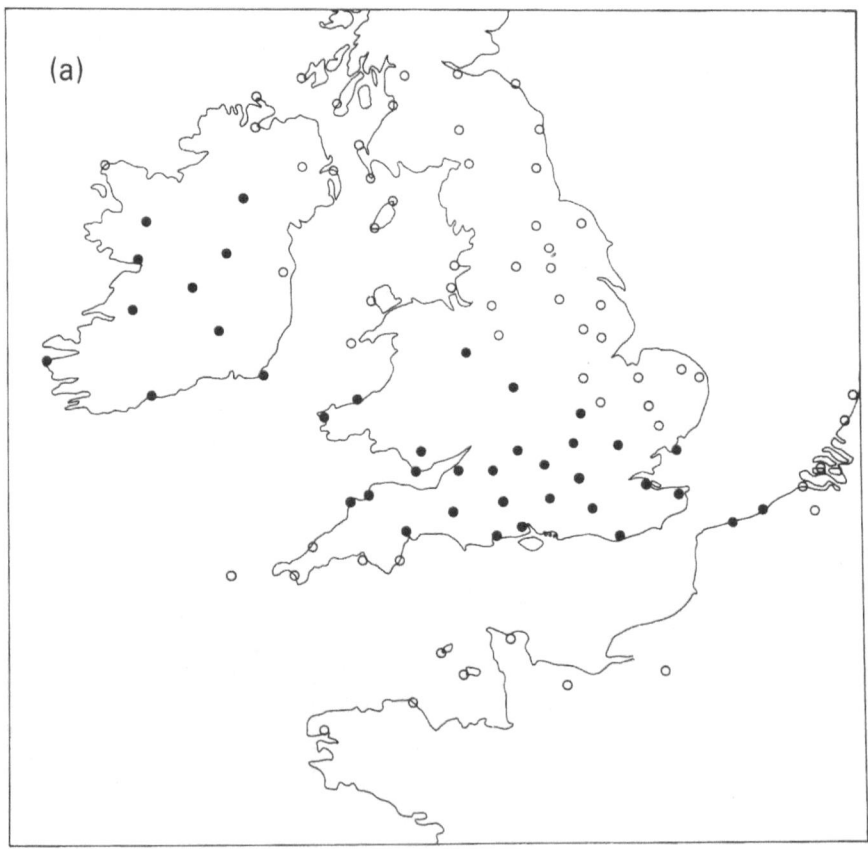

Figure 3.9 Correcting for low level evaporation of rain (from Conway and Browning
1988, with permission). (a) Surface observations of rain (•) and no rain (○) at 0300 h
GMT on 9 February 1987; (b) radar composite for the same time as (a). Rain is detected
aloft, south of The Wash, but is evaporating before reaching the ground.

Lhermitte (1988) suggested using a 94 GHz Doppler radar in conjunction with an X- or S-band radar tò determine the drop size distribution of the rain. This is based on the Mie backscattering oscillations occurring in the rain drop for drop sizes of about 1.5 mm and smaller.

Conway and Browning (1988) and Collier *et al.* (1989) illustrate how ground-based radar data from various sites, operating at 5.6 or 10 cm, are combined with conventional rain-gauge data and also with geostationary satellite data to make rainfall estimates for mesoscale meteorological forecasting. While their example is for the British Isles, the principle is generally applicable. The techniques involve human interaction as well as numerical model estimates to reduce spurious errors when the various data types are overlaid. Figure 3.9 illustrates how ground-based radar data are combined with rain-gauge data, and how rain that is evaporating before reaching the ground is removed. Similar systems are being made operational with different interactive schemes in many countries.

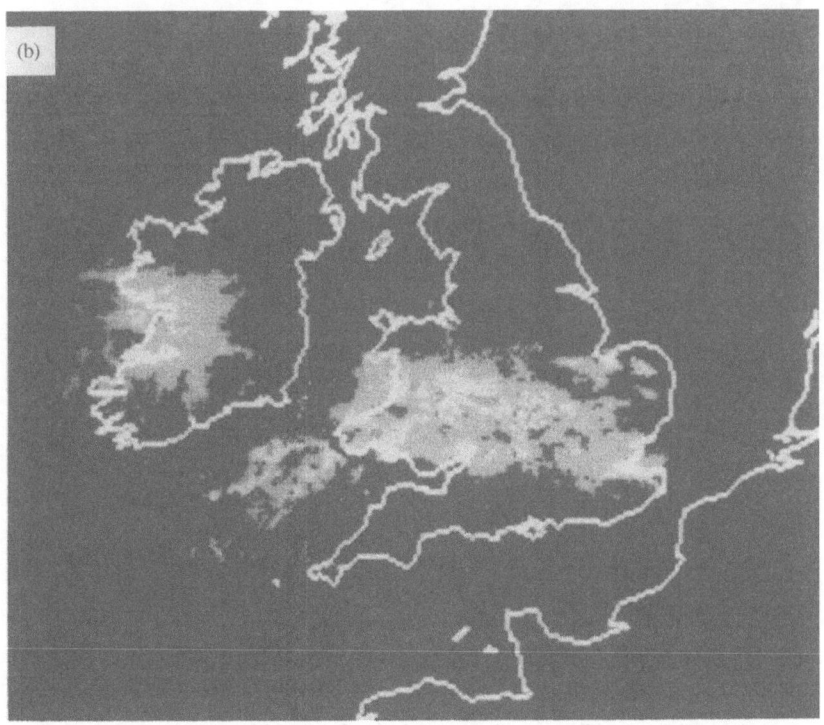

Sampling problems

In addition to the problem of estimating rain rates, there is the problem in low Earth orbit of a lack of sampling frequency. Satellites in low Earth orbit are usually in sun-synchronous orbits; that is, they pass over any point on the Earth's surface twice daily at the same time each day. Most meteorological satellites have equatorial crossing times of about 9 o'clock and 1 o'clock local time, and yet rain can fall at many other times of day. Geosynchronous orbiting satellites view the same part of the Earth at least every half-hour and can in principle monitor any point continuously in their field of observation. However, the types of observations currently capable of being made from geostationary orbit are limited, and so the effects of sampling rainfall from low

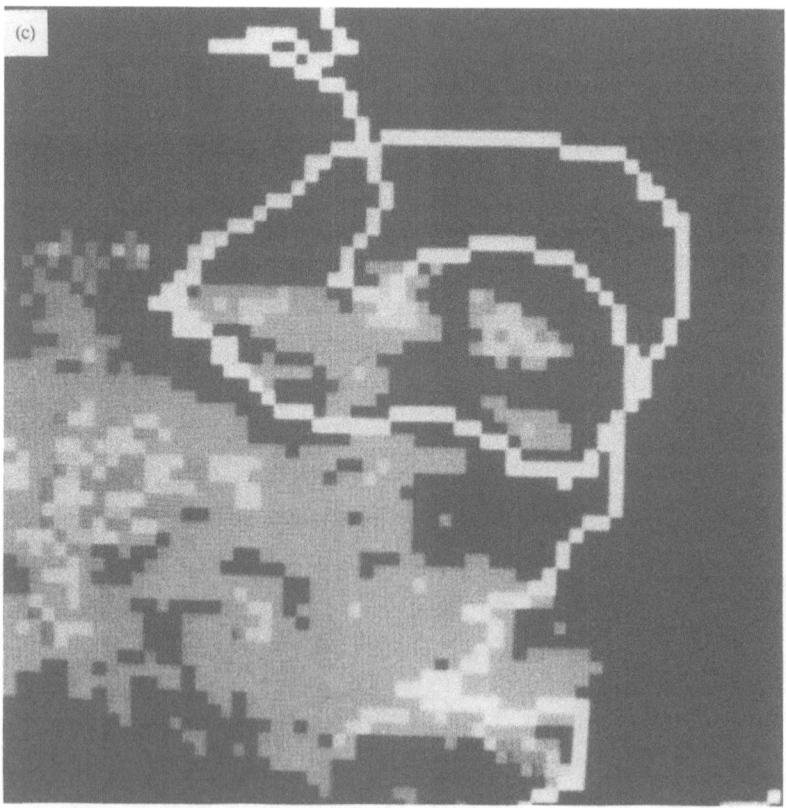

Figure 3.9 (*cont'd*)   (c) The evaporating rain is identified interactively by an operator and marked for detection; (d) the area of evaporating rain is removed.

Earth orbit must be investigated. Many of these investigations are summarized in Simpson *et al.* (1988).

Laughlin (1981) raised the sampling issue in relation to GATE (GARP (Global Atmospheric Research Program) Atlantic Tropical Experimental) data, that is rainfall data taken over the tropical Atlantic Ocean in 1974. He used a first-order Markov model to study areally averaged rain rate time series, and showed that errors would be less than 10% on monthly rainfall if a satellite observed the entire area twice daily. He also found that the auto-correlation time in the GATE area is about 6 hours, half the satellite revisit time. This order of magnitude error has been confirmed recently by several workers. These studies are summarized by North (1987), but more work needs to be done on other data sets to see if this finding is generally confirmed.

3.4 FUTURE CONSIDERATIONS

It appears likely that considerable progress will continue to be made on improving our ability to measure rainfall and rain rates from remote sensing

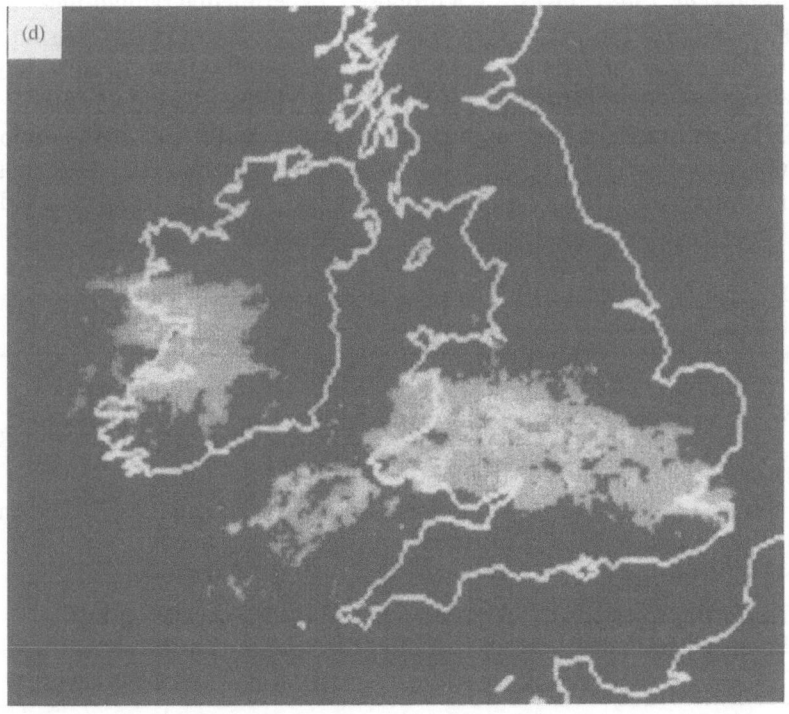

instruments. It is also expected that our capabilities for doing this from space will also develop rather rapidly, although no single approach or wavelength will be the answer. It is increasingly obvious that progress will be made using multifrequencies and multipolarizations, as well as a host of non-spaceborne instrumentation such as atmospheric soundings, ground data, etc. It also seems obvious that, owing to the inexact nature of the science, the use of interactive computers and expert systems will be incorporated to assist analysis in the more subjective aspects of estimating rainfall.

It also appears likely that hierarchical operational approaches will be developed and perfected. Such approaches may use an objective VIS/IR method for estimating large-area long-term (5 days or more) rain totals. Within this large framework, interactive techniques such as BIAS with or without passive microwave data would be used to analyse significant events and frontal storms. Embedded further in the system might be an interactive severe storm analysis procedure for predicting high-intensity rain rates and making flash flood forecasts. Such a hierarchical system could be developed for relatively inexpensive personal computers (PCs) so that the entire system could be located in the field where the information is needed.

There are also a number of precipitation-related satellite programmes on the drawing board. One of these is TRMM (Tropical Rainfall Measuring Mission) (Simpson *et al.* 1988). TRMM is a proposed, experimental rainfall-measuring satellite designed for a minimum 3 year mission to measure the distribution and variability of rainfall and latent heat release over tropical and subtropical regions of the globe. The purpose of TRMM will be to improve short-term climatological models, general circulation models and the global hydrologic balance, particularly as it is affected by tropical oceanic rainfall. The details of the instrumentation and the orbit are listed in Table 3.5. It is expected to

Table 3.5 TRMM sensor summary (from Simpson *et al.* 1988). Orbit 30°–35° inclination, 320 km altitude (rapid precession)

Microwave radiometers	*Radar*	*Visible/infrared radiometer*
19, 37, 90 GHz	14 and 24 GHz	
dual polarized	4 km footprint	VIS and 10 μm IR at 1 km
at 10 km resolution	250 m range resolution	resolution
600 km swath	220 km swath	600 km swath
10 GHz at 20 km resolution★	600 km swath★	Moonlight visible★
5 GHz at 40 km resolution★		1.6 μm (phase)★
		6.7 μm, split window★

★ Items desired if resources permit but are not necessary to achieve main TRMM objectives.

clarify problems in the VIS/IR/PMW data interpretation areas and should therefore have a positive impact on the existing types of satellite techniques discussed earlier in this chapter.

Another future programme that will undoubtedly have great potential for estimating precipitation will be the Earth observing system (Eos) being proposed by NASA (NASA 1984). Eos will be a multiplatform system of sensors dedicated both to monitoring environmental conditions and to measuring Earth resources. There are several instruments that have been selected and which will be valuable for estimating rainfall and rain rates.

REFERENCES

Atlas, D. and Thiele, O. W. (1981) *Precipitation Measurements from Space*. Workshop Report NASA, Goddard Space Flight Center, Greenbelt, MD.

Barrett, E. C. (1981) AgRISTARS state I: Development and application of the Bristol method for improved rainfall monitoring for mid-latitudinal use. Final Report to US Dept of Commerce, Washington DC, Remote Sensing Unit, University of Bristol.

Barrett, E. C., Beaumont, M. J., Harrison, A. and Richards, T. S. (1986) BIAS I²S users guide. Final Report to US Dept of Commerce, Washington DC, Remote Sensing Unit, University of Bristol.

Barrett, E. C., D'Souza, G., Power, C. H. and Kidd, C. (1988) Towards trispectral satellite rainfall monitoring algorithms. In *Tropical Precipitation Measurements, Proc. Int. Symp., Tokyo, Japan, NASA/NASDA* (eds J. S. Theon and J. Fugono), A.DEpak, Hampton, VA, pp. 285–92.

Barrett, E. C. and Kidd, C. (1987) The use of SMMR data in support of a VIR/IR satellite rainfall monitoring technique in highly-contrasting climatic environments. In *Passive Microwave Observing from Environmental Satellites, a Status Report* (ed. J. C. Fischer), *NOAA Tech. Rep. NESDIS* 35, Washington DC, pp. 109–23.

Barrett, E. C. and Martin, D. W. (1981) *The Use of Satellite Data in Rainfall Monitoring*, Academic Press, London.

Clark, D. and Morris, D. G. (1986) The NOAA satellite precipitation estimate program — with preliminary results of an operational test using estimates in a hydrologic river forecast model. *Hydrologic Applications of Space Technology (Proc. Cocoa Beach Workshop, Florida, August 1985), IAHS Publ. No.* 160, pp. 41–6.

Collier, C. G., Goddard, D. M. and Conway, B. J. (1989) Real-time analysis of precipitation using satellites, ground-based radars, conventional observations and numerical model output, *Meteorol. Mag.* 118, 1–8.

Conway, B. J. and Browning, K. A. (1988) Weather forecasting by interactive analysis of radar and satellite imagery. *Philos. Trans. R. Soc.* A 324, 299–315.

Dahlstrom, B. (1985) Estimation of precipitation by remote sensing techniques. In *Remote Sensing Applications to Hydrology and Water Resources, Proc. Int. Semin. organized by the Slovak Hydrometeorological Institute and Datasystem K.U.O., UNESCO Technical Documents in Hydrology*, pp. 46–56.

D'Souza, G. and Barrett, E. C. (1988) A comparative study of candidate techniques for U.S. heavy rainfall monitoring operations using meteorological satellite data. Final

Report to US Dept. of Commerce, Washington DC, Remote Sensing Unit, University of Bristol.

Ferraro, R. J., Grody, N. C. and Kogut, J. A. (1986) Classification of geophysical parameters using passive microwave satellite measurements. *IEEE Trans. Geosci. Remote Sensing* **GE-24**, 1008–13.

Goldhirsh, J. (1988) Analysis of algorithms for the retrieval of rain-rate profiles from spaceborne dual-wavelength radar. *IEEE Trans. Geosci. Remote Sensing* **GE-26**, 98–114.

Goldhirsh, J., Rowland, J. and Musiani, B. (1987) Rain measurement results derived from a two-polarization frequency-diversity S-band radar at Wallops Island, Virginia. *IEEE Trans. Geosci. Remote Sensing* **GE-25**, 654–61.

Grody, N. C. (1984) Precipitation monitoring over land from satellite by microwave radiometry. *Int. Geoscience and Remote Sensing Symp. (IGARSS'84), Strasbourg, France*, ESA SP-215, pp. 417–23.

Griffith, C. G., Woodley, W. L., Grube, P. G., Martin, D. W., Stout, J. and Sikdar, D. (1978) Rain estimation from geosynchronous satellite imagery — Visible and near infrared studies. *Mon. Weather Rev.* **106**, 1153–71.

Laughlin, C. R. (1981) On the effect of temporal sampling on the observation of mean rainfall, precipitation measurements from space, D55–D66. *Workshop Rep.* (eds D. Atlas and O. W. Thiele), NASA Goddard Space Flight Center, Greenbelt, MD.

Lhermitte, R. M. (1988) Cloud and precipitation remote sensing at 94 GHz. *IEEE Trans. Geosci. Remote Sensing* **GE-26**, 207–16.

Linsley, R. K. and Franzini, J. B. (1955) *Elements of Hydraulic Engineering*, McGraw-Hill, New York.

Meneely, J. M. (1975) *Applications of the Nimbus-5 ESMR to rainfall detection over land surfaces. NASA Contractor Report NAS5-20878*. Goddard Space Flight Center, Greenbelt, MD.

Meneghini, R., Eckerman, J. and Atlas, D. (1983) Determining rain rate from a spaceborne radar using measurements of total attenuation. *IEEE Trans. Geosci. Remote Sensing* **GE-21**, 34–43.

Moses, J. F. and Barrett, E. C. (1986) Interactive procedures for estimating precipitation from satellite imagery. *Hydrologic Applications of Space Technology (Proc. Cocoa Beach Workshop, Florida, August 1985)*, IAHS Publ. No. 160, pp. 25–39.

NASA (1984) *Earth Observing System*, vols I and II, NASA TM-86129, NASA/ Goddard Space Flight Center, Greenbelt, MD.

NASA (1987) *High Resolution Multichannel Microwave Radiometer, Eath Observing System*, vol. IIe, NASA/Goddard Space Flight Center, Greenbelt, MD.

North, G. R. (1987) Sampling studies for satellite estimation of rain. *Proc. 10th Conf. on Probability and Statistics in Atmospheric Science, Edmonton, Canada*, American Meteorological Society, Boston, MA.

Savage, R. C. and Weinman, J. A. (1975) Preliminary calculations of the upwelling radiance from rain clouds at 31.0 and 19.5 GHz. *Bull. Amer. Meteor. Soc.* **56**, 1272–4.

Schultz, G. A. (1989) Remote sensing of watershed characteristics and rainfall input. In *Unsaturated Flow in Hydrologic Modeling-Theory and Practice, NATO ASI Ser., Ser. C*, vol. 275 (ed. H. J. Morel-Seytoux), Kluwer, Dordrecht, pp. 301–23.

Schultz, G. A. and Klatt, P. (1980) Use of data from remote sensing sources for hydrological forecasting. *Hydrological Forecasting — Proc. Oxford Symp., IAHS Publ. No.* 128, pp. 75–82.

Scofield, R. A. (1986) Satellite convective and extratropical cyclone cloud categories associated with heavy precipitation. *Hydrologic Applications of Space Technology (Proc. Cocoa Beach Workshop, Florida, August 1985), IAHS Publ. No.* 160, pp. 47–57.

Scofield, R. A. and Oliver, V. J. (1977) A scheme for estimating convective rain from satellite imagery. *NOAA/NESS Tech. Memo.* 86, US Dept of Commerce, NOAA, Washington DC.

Scofield, R. A. and Oliver, V. J. (1981) A satellite derived technique for estimating rainfall from thunderstorms and hurricanes. *Satellite Hydrology*, American Water Resources Association, Minneapolis, MN, pp. 70–6.

Simpson, J., Adler, R. F. and North, G. R. (1988) A proposed tropical rainfall measuring mission (TRMM) satellite. *Bull. Am. Meteorol. Soc.* **69**, 278–95.

Spencer, R. W., Goodman, H. M. and Wood, R. E. (1988) Precipitation retrieval over land and ocean with the SSM/I, part 1: Identification and characteristics of the scattering signal. *J. Atmos. Oceanic Tech.* **2**, 254–63.

Ulbrich, C. W. (1986) A review of the differential reflectivity technique of measuring rainfall. *IEEE Trans. Geosci. Remote Sensing* **GE-24**, 955–65.

Weinman, J. A. and Gunther, P. J. (1977) Determination of rainfall distributions from microwave radiation: measured by Nimbus-6 ESMR. *J. Appl. Meteor.* **16**, 437–42.

Wilheit, T. T., Rao, M. S. V., Chang, T. C., Rogers, E. B. and Theon, J. S. (1977) A satellite technique for quantitative mapping rainfall rates over the ocean. *J. Appl. Meteorol.* **16** (S), 551.

4

Snow hydrology

4.1 INTRODUCTION

Snow is a form of precipitation, but, in hydrology it is treated somewhat differently because of the lag between when it falls and when it produces runoff, groundwater recharge, and is involved in other hydrologic processes. The hydrologic interest in snow is mostly in mid- to higher latitudes and in mountainous areas where a seasonal accumulation of a snowpack is followed by an often lengthy melt period that sometimes lasts months. During the accumulation period there is usually little or no snowmelt. Precipitation falling as snow (and sometimes rain) is temporarily stored in the snowpack until the melt season begins. The hydrologist generally wants to know how much water is stored in a basin in the form of snow. The hydrologist will also be concerned with the areal distribution of the snow, its condition and the presence of liquid water in it. In general, all these indicators of snow are difficult to measure and are likely to vary considerably from point to point, especially in mountainous terrain.

Remote sensing offers a new and valuable tool for obtaining snow data for predicting snowmelt runoff. Historically, snow data have been obtained manually by means of snow courses, which are extremely labour intensive, expensive and potentially dangerous. Even when available, snow course data represent only a point and, at best, can only be used as an index of available snow water content. Recent use of telemetering of snow pillow and storage gauge measurements of precipitation have reduced the need for some fieldwork but have not overcome the problem of the point measurements of snow. That is, although measurements are automated, there is still the problem that a single point measurement may or may not be representative of a large area or basin. From a remote sensing perspective, snow cover is one of the most readily identifiable measures of water resources from aerial photography or satellite imagery. Present operational satellite systems are limited to determining only the area of snow cover; depth or snow water equivalent cannot be measured directly by these systems. However, it will be shown in the following sections that there is a capability for estimating these now from research instruments, and it appears that future operational systems will also provide very important snow hydrology data. A more complete description of remote sensing of snow can be found in Hall and Martinec (1985).

4.2 GENERAL APPROACH

Just about all regions of the electromagnetic spectrum can provide useful information about the snowpack and its condition. Ideally, we would like to know the areal extent of the snow, its water equivalent, and the 'condition' or grain size, density and presence of liquid water. Although one region of the spectrum can provide all these properties, certain regions of the spectrum can be used to measure individual properties. Each of the remote sensing approaches will be addressed below, starting with the technique at the shortest wavelength. Table 4.1 (Rango 1983) summarizes some of the sensor responses to snowpack properties.

Gamma radiation

The water content of some snowpacks can be measured with low-flying aircraft carrying sensitive gamma radiation detectors. This method takes advantage of the natural emission of low-level gamma radiation from the soil. Naturally occurring radioisotopes of potassium, uranium and thallium can be found in a typical soil. Aircraft passes over the same flight line before and during snow cover measure the attenuation resulting from the snow layer which is empirically related to an average snow water equivalent for that site (Carroll and Vadnais 1980). This approach is limited to low aircraft altitudes (approximately 150 m) because the atmosphere attenuates a significant portion of the radiant energy. This restriction effectively limits the use of gamma detection to relatively flat areas, and because of safety considerations it cannot be used in mountainous areas. Also, this approach has been limited for the most part to non-forested areas because the effect of forest biomass is to attenuate the radiation signal (Glynn et al. 1988). However, recent work by Carroll and Carroll (1989) suggests a means of correcting for the downward bias in measured snow water equivalent by making a correction that is based on the amount of biomass and the type of radiation. In addition, because the

Table 4.1 Sensor band responses relative to various snowpack properties (Rango 1983)

Property	Sensor band: visible/near infrared	Thermal infrared	Microwave
Snow-covered area	Yes	Yes	Yes
Depth	If very shallow	Weak	Moderate
Snow water equivalent	If very shallow	Weak	Strong
Stratigraphy	No	Weak	Strong
Albedo	Strong	No	No
Liquid water content	Weak	Weak	Strong
Temperature	No	Strong	Weak
Snow/soil boundary	No	No	Weak (high frequency to strong low frequency)
All weather capability	No	No	Yes
Current best spatial resolution from space platform	Tens of metres	Hundreds of metres	Passive: 30 (high frequency) to 150 km (low frequency); active: tens of metres

gamma energy is relatively low, the maximum depth of snow water equivalent is limited to about 30–40 mm and interpretation of the data can be difficult if the background soil moisture changes during the season (Vershinina 1985). However, Carroll and Vose (1984) have reported measurements in a forest environment with a snow water equivalent of 480 mm. In the NOAA operational airborne gamma radiation snow water mapping programme, procedures for correcting for the soil moisture are included in the system (Carroll and Carroll 1989). Currently, this operational programme flies over 1400 flight lines in the United States and Canada (Carroll and Carroll 1989).

Visible/near infrared

The albedo of the snow surface is the property most easily measured by remote sensing. Albedo, A, is defined as

$$A = \text{Reflected solar radiation/Incoming solar radiation} \qquad (4.1)$$

Typically, new snow will have an albedo of 90% or more whereas older snow that has been weathered and has accumulated dust and litter can have an albedo as low as 40% (Foster *et al* 1987). The reflectivity depends upon snow properties such as the grain size and shape, water content, surface roughness,

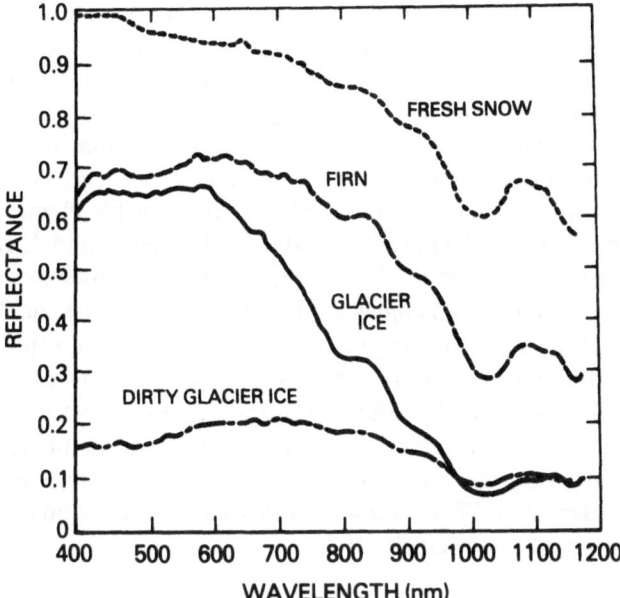

Figure 4.1   An illustration of the spectral reflectance curves of snow and ice (after Hall and Martinec 1985).

depth and presence of impurities. The reflectivity of new snow decreases as it ages in both the visible and infrared regions of the spectrum; however, the decrease is more pronounced in the infrared region as shown in Figure 4.1. This increased sensitivity in the infrared region is caused by the increasing grain size of the snow which results from melting and refreezing. For the most part, decreased reflectivity in the visible region can be attributed to contaminants such as dust, pollen and aerosols.

Most current snowmelt models use either point snow course estimates of the snowpack or the total area of the snowpack. In the latter case there is usually an implicit assumption that the snow cover area and its changes are somehow consistently related to the snow water equivalent in the basin. Through careful analysis of satellite images or aerial photography, snow can be identified and the boundaries of the snow/no-snow areas accurately located. However, a certain amount of subjective interpretation may be necessary to identify and separate the effects of shadows and forests.

In non-forested terrain, all areas with continuous brightness distinctly greater than the normal dark background, and that have been assured to be cloud free, should be mapped as snow areas. The snow line that encloses these areas can be assumed to represent accumulated snow depths of 2.5 cm (Bowley *et al.* 1981) or more. Areas on imagery that appear mottled (alternating dark and light reflectance) can be mapped as areas of between zero and 2.5 cm of snow depth. In mountainous terrain, the snow line is mapped at the edge of the brighter tone without regard to brightness variations resulting from forest effects or mountain shadow.

Snow can readily be identified and mapped with the visible bands of satellite imagery because of its high reflectance in comparison with no-snow areas. Generally this means selecting the NOAA VHRR visible channel, Landsat MSS channels 4 or 5, SPOT, or Landsat TM channels 2 and 4. If there is a choice of bands it is better to choose a spectral band closer to the infrared region because at higher sun angles the Landsat MSS band 4 and TM band 1 and other bands near the blue region may be near saturation, causing a loss of detail in identifying snow and no-snow areas. Although snow can be detected at longer wavelengths, that is in the near infrared, the contrast between a snow and a no-snow area is considerably lower than with the visible region of the spectrum. However, the contrast between clouds and snow is greater in Landsat TM Band 5 (1.57–1.78 μm) and this serves as a useful discriminator between clouds and snow (Dozier 1984).

Thermal infrared

Thermal data are perhaps the least useful of the common remote sensing products for measuring snow and its properties. In order to determine snowpack temperatures the spectral emissivity of the snow must be known. This in turn requires knowledge of liquid water content and the grain size as well as other factors. In spite of these limitations, thermal data can be useful for helping identify snow/no-snow boundaries. Thermal infrared data are also useful for discriminating between clouds and snow with AVHRR data because the 1.57–1.78 μm band is not available on this sensor.

The emissivity of snow approaches that of a black body at the 10.5–12.5 μm atmospheric window. Griggs (1968) has shown that melting snow can have an emissivity as high as 99%, whereas the emissivity for no-snow areas is typically 95% or less.

Microwave

Microwave remote sensing offers great promise for future applications to snow hydrology. This is because the microwave data can provide information on the snowpack properties of most interest to hydrologists; that is, snow cover area, snow water equivalent (or depth) and the presence of liquid water in the snowpack which signals the onset of melt (Kunzi et al. 1982). The microwave potential for snow is based on the sensitivity to the presence of even minute quantities of liquid water in the snow which results in drastically different dielectric constants of the snowpack. The emission and the reflection of microwave radiation exhibit the ability to penetrate not only clouds and most precipitation, but also the snowpack itself, at least under certain conditions. However, microwave interaction with snow is extremely complex because snow properties such as depth, dryness, crystal size, liquid water content and the underlying soil all affect the response. Active microwave systems offer extremely good spatial resolution from space platforms at the expense of more complexity in the analysis and interpretation of the data. On the other hand, passive systems may result in more simple data interpretation at the expense of relatively poor spatial resolution.

4.3 SNOW MAPPING

The aerial extent of snow cover can be determined through various remote sensing techniques. Snow cover area by itself is of little value for predicting snowmelt runoff because it provides no information on the depth or water equivalent. However, useful empirical relationships can be developed for specific basins and the temporal nature of some remote sensing data allow the inference of the rate of melt. Leaf (1969) used aerial photographs to develop relationships between snow cover and accumulated runoff for some Colorado watersheds. Sequential photographs showing snow cover depletion were used to help estimate the timing and magnitude of snowmelt peaks.

Since 1973, the NOAA and Landsat satellites have provided visible and infrared imagery of snow cover throughout the world. Procedures have been developed for analysing these data to determine snow cover area (Meier and Evans 1975; Rango and Itten 1976).

Use of satellite data for snow mapping has become operational in several regions of the world. Currently, NOAA develops daily snow cover maps for 56 river basins in the USA for use in streamflow forecasting (Carroll and Allen 1988). NOAA also produces maps of mean monthly snow cover for the northern hemisphere (Wiesnet and Matson 1975; Matson and Wiesnet 1981) using the lower resolution (4–8 km) visible data. The Meteor satellite has been used to delineate snow/no-snow lines for river basins and other areas in the USSR. Meteor and NOAA data were combined to map snow cover area on basins ranging from 530 to more than 12 000 km^2 (Shcheglova and Chernov 1982). A handbook to assist potential users of satellite data in the mapping of snow-covered areas has been published by Bowley et al. (1981).

The NOAA data have a distinct advantage for mapping snow over the higher-resolution Landsat and SPOT satellites in the frequency of observation. Cloud cover can negate the usefulness of a Landsat overpass that occurs only once every 16 days. The pointing capability of SPOT increases its flexibility somewhat but the user is still limited to certain day windows of observation. For most hydrologic applications, the compromise made in accepting the coarser spatial resolution as the price to pay for more frequent observations is not serious. With 1 km NOAA data, Odegaard *et al.* (1979) have reported using digital techniques on basins as small as 20 km^2. For smaller basins, Landsat or SPOT data would have to be used, or, as an alternative, low-level aircraft flights. The Landsat MSS (80 m) data can be used on basins as small as 10 km^2 and the SPOT or Landsat TM data should be able to be used in basins as small as 2–5 km^2 (Rango 1983).

Snow mapping procedures

There are several approaches that can be taken for mapping snow, depending upon the data used, the equipment available and the proposed use of the maps. The documented procedures most commonly used are described below.

1. *Zoom transfer scope.* A zoom transfer scope can be used for manually transferring data representing a snow/no-snow line directly on to a base map. With the zoom transfer scope, the satellite image (either a print or a transparency) is optically superimposed over the base map. The image can be adjusted by stretching to coincide exactly to the base map scale. The operator then subjectively delineates the snow line directly on the base map.
2. *Manual transfer.* This procedure requires developing coarse and fine grids of the same scale as the satellite imagery to transfer the snow line directly on to basin contour maps. The grids are prepared on transparent overlays so that the snow boundaries can be traced and then transferred manually to the corresponding topographic map. It is suggested that a scale of 1 : 250 000 and a coarse grid of 30 s be used for most applications. A colour-additive viewer can also be used to produce either natural colour composites or false-colour infrared composites from which the snow-covered area can be traced on to a transparent overlay.
3. *Density slicing.* Density slicing uses a positive satellite image with all but the basin masked out. A camera records the various shades of grey into a number of discrete levels which can be displayed on a CRT monitor in false colour. The operator subjectively decides which colours represent snow-covered areas and the machine electronically planimeters these colours to determine the total area of snow cover.
4. *Digital analysis.* Interactive image analysis systems and computer-aided classification of snow/no-snow areas are now being used operationally. The following are the steps used by the US National Weather Service (Carroll and Allen 1988) in the operational digital analysis of GOES data:
   (a) identify areas to be mapped and select data sectors;
   (b) develop an automated system to collect full resolution data;
   (c) set up 7-day rotating files to archive the data;
   (d) input US Geological Survey (USGS) cataloguing units for basin delineation over areas to be mapped;

(e) prepare snow-free/cloud-free reference databases (masks);
(f) remap GOES data into mask projection;
(g) align, pixel for pixel, the mask and GOES data;
(h) adjust the brightness of the mask to reflect the difference in solar illumination angle of the GOES data;
(i) composite two images to eliminate clouds;
(j) compare each pixel in the mask and image to produce a snow/no-snow image;
(k) eliminate cloud contamination manually with the cursor;
(l) overlay basin boundaries and digitally compute percentage of snow cover;
(m) overlay percentage of snow cover tables for each basin;
(n) archive GOES data used for mapping and snow/no-snow digital data sets on magnetic tape.

Lucas *et al.* (1989) at the Remote Sensing Unit in Bristol, UK, have developed an unsupervised multispectral classification for discriminating between snow and clouds and for locating peripheral areas of the snowpack that are undergoing melt. The system uses AVHRR data and has supporting software for correction of forest effects and cloud shadows. The NOAA-9 AVHRR channels 1, 3 and 4 are used to remove cloud areas with new snow, whereas channel 2 minus channel 1, together with channels 3 and 4, are used during periods of advanced snowmelt (Harrison and Lucas 1989). Frank *et al.* (1988) have used NOAA AVHRR channels 3, 4 and 5 to produce a satisfactory separation of clouds and snow for areas of the Swiss Alps.

Confusion factors

When mapping snow cover there are several possible features of the imagery which must be considered in order to prevent misidentification of snow or no-snow areas. With experience, techniques have been developed to overcome some of the possibilities of misinterpretation because of the following confusion factors.

1. *Clouds*. Cloud tops exhibit a very bright reflectance in the visible bands that is often indistinguishable from snow. Differentiating between clouds and snow is one of the major problems in the use of satellite data for snow mapping. An experienced user will use some or a combination of terrain features, pattern recognition, uniformity of reflectances, the presence of shadows (either from terrain features or clouds) and scene stability with time. Snow can be distinguished from clouds by using a near-infrared channel around 1.6 μm because the cloud reflection will be bright in this region but the snow will be dark (Crane and Anderson 1984; Dozier 1984).

2. *Forest cover*. Forested areas can consist of everything from dense conifers to less dense deciduous forests, to sparse range-type vegetation. The reflectance from these areas will be considerably darker than non-forested areas even with substantial depths of snow because the snow will tend to filter through the forest canopy. The challenge is to determine the snow-covered areas when they may not be directly detectable. This generally requires a great deal of experience and familiarity with the area and the use of all concomitant information available (i.e. land use surveys, topographic maps, non-snow imagery, etc.). Timber cuts, roads, streams, lakes and other open land can be used because they would be highly reflective under snow cover. Digital enhancement of data can also sometimes be used if the forest cover is incomplete or sparse.
3. *Shadows*. During the winter, sun angles are generally low and the resulting northern hemisphere terrain shadows on north-facing slopes may be difficult to distinguish from bare, south-facing slopes. Topographic maps and summer imagery may help in the interpretation. In shadow areas, snow may be distinguished from rocks or soil by selecting a threshold brightness for automated discrimination (Dozier and Marks 1987).
4. *Rocks*. During the melt period, highly reflecting bare rock may be difficult to distinguish from late season snow. As above, summer imagery and topographic maps, as well as vegetation patterns, can help in the differentiation.

4.4   CURRENT APPLICATIONS

Different approaches for determining snow area, water equivalent and snow properties have been developed. These have been driven for the most part by the availability of data from existing satellites or from experimental aircraft and truck programmes. This section discusses a number of examples, some of which are operational and others still considered experimental or developmental.

Snow cover area

Snow cover area is not the ideal description of a snowpack. A hydrologist would like to have water equivalent, depth and density information, as well as the snow cover area, to make accurate estimates of snowmelt runoff. Fortunately, research has demonstrated a very good relationship between runoff and snow cover area for many basins.

In an early application of satellite data, Rango *et al.* (1977) used simple photointerpretation techniques to map snow cover areas in the Indus and Kabul river basins in Pakistan. Their approach was to use a simple regression between the percentage of snow cover in the basins from 1 to 20 April and the

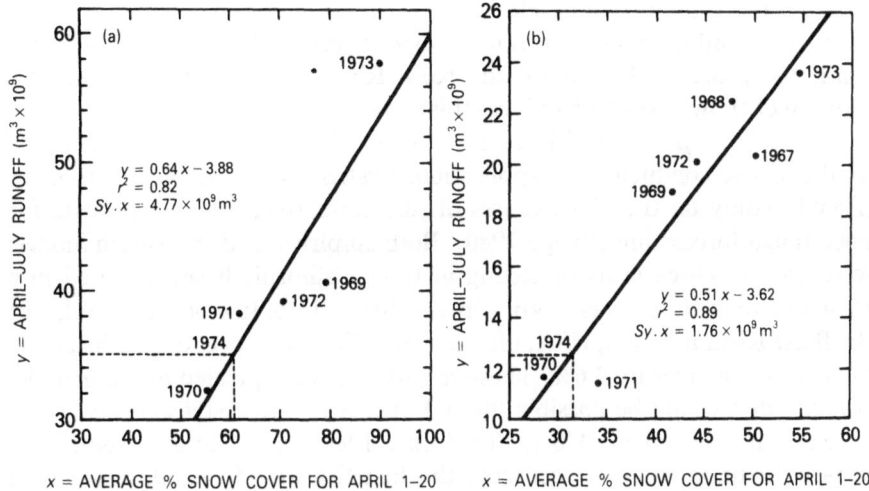

Figure 4.2 Illustrations of satellite-derived snow cover estimates against measured runoff for the Indus River above Besham, Pakistan, 1969–1973 (a) and for the Kabul River above Nowshera, Pakistan, 1967–1973 (b). Seasonal runoff for 1974 is predicted on the basis of actual 1974 snow cover measurements (after Rango *et al*. 1977).

April–July streamflow. Figure 4.2 illustrates the relationship and the resulting R^2 values. These results demonstrated the usefulness of satellite-derived runoff estimates, especially for remote and data-sparse regions of the world. This work was extended (Dey *et al*. 1983) with an additional 6 years of data. The additional data improved the regression for the Kabul River but decreased it for the Indus River. Such results point out the inherent weakness of simple regression models representing complex processes. On the other hand, in data-sparse regions there are seldom many alternative approaches.

Landsat imagery was used to determine snow cover areas for six basins in Colorado, USA, over the period of 1973 to 1978. Shafer and Leaf (1979) concluded that the satellite imagery was of sufficient quality to monitor the snow cover area accurately. They also concluded that forecast error can be reduced by the order of 10% by using snow cover data derived from the satellite.

Aircraft and Landsat snow cover data were combined to form a long-term database for predicting runoff for basins in California. Two areas were studied by comparing snow cover areas with conventional snow data and by incorporating snow cover areas into the State's forecasts (Brown *et al*. 1979). The results indicated a potential improvement in forecast accuracy by using snow cover area, particularly in areas where precipitation and snow course data are limited. The Kings River basin is much more predictable, in the sense of having a more uniform area–elevation distribution of snow, than the Kern River basin and its standard estimated error is relatively small. On the other hand, the procedural error for the Kern River is relatively large and the addition of snow cover data reduces this error.

NASA, in cooperation with several federal and state water resource agencies in the USA, conducted an applications systems verification and transfer (ASVT) study on the effectiveness of satellite-derived snow cover data for operational forecasting (Rango 1980). Both empirical and short-term models were tested. Three years of testing in three California basins resulted in a reduction in the forecast errors of between 10 and 15%. In modelling studies of the Boise River in Idaho, USA, the use of satellite snow cover data reduced the 5-day forecast error by 9.6%. These results were extrapolated to estimate the benefits that would be possible through increased forecast accuracies in the 11 western states in the USA. Based on a 1980 dollar and an assumed 6% improvement in forecast accuracy, the benefits would include more than $10 million dollars from improved hydropower predictions and $28 million in irrigation water forecasting. A benefit/cost ratio was calculated to be an impressive 75 : 1 (Castruccio *et al.* 1980). However, this figure did not consider the cost of satellite development and launch, and it also did not anticipate any improvement in satellites, models or interpretation systems.

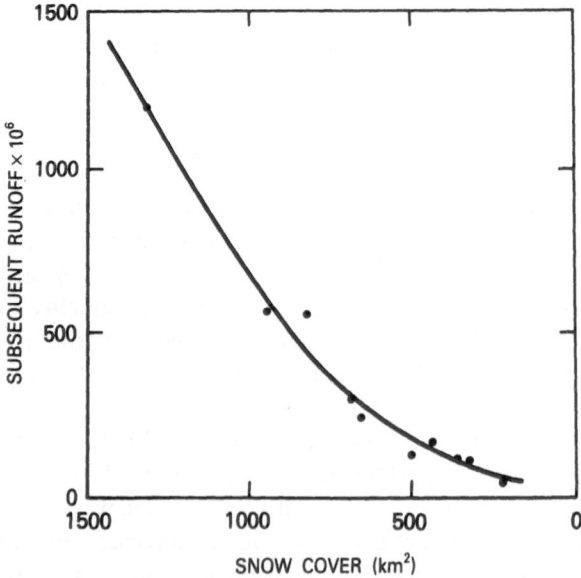

Figure 4.3   An illustration of the relation between remaining snow as of a given date and the subsequent melt water runoff (m^3) (after Ostrem *et al.* 1981).

Satellite data have been used for determining snow cover area in Norway for forecasting snowmelt runoff and managing the production of hydroelectric power (Ostrem et al. 1981). These researchers have developed a method for using NOAA and TIROS data for measuring the remaining snow and to predict the corresponding snowmelt runoff volume for a number of Norwegian high mountain basins. The method is limited to essentially vegetation-free areas and only after roughly 20% of the basin is snow free. A relationship similar to that shown in Figure 4.3 is used to predict the subsequent snowmelt runoff from the satellite-determined snow cover area. Such results, although empirical, can be developed for many basins.

The question of spatial resolution of remotely sensed data necessary to achieve satisfactory results when mapping snow cover has been addressed by Rango et al. (1983). Their results are summarized in Table 4.2. Since their study, SPOT data have become available, and we might expect the usefulness of SPOT to fall between the orthophoto and Landsat TM.

Passive microwave techniques can be used to identify snow cover under certain circumstances. When the snow is dry, the emissivity at microwave frequencies is relatively low and snow–soil boundaries can be identified with measurements at 18 or 37 GHz. Burke et al. (1984) reported that the 37 GHz microwave data displayed a decrease in brightness temperature as depth increased, but at 18 GHz depth was not a significant factor.

Rango et al. (1979) reported data that showed a fairly good inverse relationship between snow accumulation and brightness temperature from ESMR-6 (Figure 4.4) for a relatively deep snow site in North Dakota, USA. However, a similar analysis for an Indiana site in which the snow cover never

Table 4.2 Characteristics of various remote sensing data used for snow cover mapping (Rango et al. 1983)

Platform sensor/data	Nominal resolution (visible)	Minimum basin size (digital/photo)	Repeat period
Aircraft			
orthophoto	3 m	1 km^2	As needed
Landsat			
TM	28.5 m	2.5/5 km^2	16 days
RBV	40 m	5/10 km^2	18 days
MSS	57 m	10/20 km^2	16 days
NOAA			
AVHRR	1.1 km	200/500 km^2	12 h
GOES			
VISSR	1.1 km	200/500 km^2	As needed

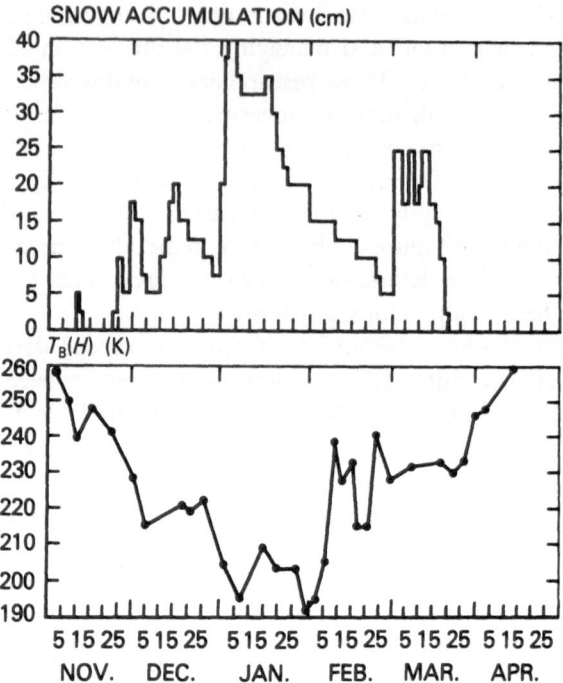

Figure 4.4 An illustration of the relationship between snow accumulation and Nimbus-6 ESMR horizontally polarized brightness temperature data (1975–1976) for Williston, North Dakota (after Foster *et al.* 1984).

exceeded 12.5 cm was not nearly as strong. Thus, it appears that shallow snow cover may not be detectable with passive microwave but that there may be a threshold above which these systems can determine snow cover and possibly the actual depth. Also, once the snow begins to melt the emissivity of melting snow and bare ground are very similar.

The algorithm for the snow/no-snow determinations are rather simple depending upon the type of data available. Patil *et al.* (1981) developed a simple algorithm for the Nimbus-7 satellite based on the simple difference of brightness temperatures at 18 and 37 GHz. Rango *et al.* (1979) reported an analogous differencing algorithm based on the horizontally polarized brightness temperature and the vertically polarized brightness temperature. Both approaches depend on the difference exceeding an empirically determined threshold to indicate snow cover.

Snow depth

In addition to the snow cover area, remote sensing techniques can provide additional information about the snowpack. Snow depth can in some cases be inferred from visible data if the snow depth is less than about 30 cm (McGinnis *et al.* 1975), but not at greater depths. McGinnis *et al.* (1975) have correlated differences in albedo determined from the NOAA satellite with snow depth in the south-eastern USA. However, this approach is limited to relatively shallow snowpacks.

Recent passive microwave research based on theoretical concepts (Hallikainen 1984; Chang *et al.* 1987) has led to simple snow depth algorithms. Chang *et al.* (1987) assumed a snow density of 0.30 and a grain size of 0.35 mm to develop the following algorithm:

$$SD = 1.59[T_B(18H) - T_B(37H)] \qquad (4.2)$$

where *SD* is the snow depth in centimetres, and $T_B(18H)$ and $T_B(37H)$ are the brightness temperatures for the SMMR 18 and 37 GHz horizontal polarization channels respectively. Plate 2 shows the snow cover area of the northern hemisphere for February 1985 using the Nimbus-7 data.

Snow water equivalent and other snow properties

Determination of density changes of snow and the presence of liquid water in the snow are very important to hydrologists because they are a common signal of incipient melt. The presence of liquid water in the snow *per se* does not change the spectral reflectance of snow. However, the process of metamorphosis that occurs during the snow season and is accelerated as the melt season approaches does have an effect on the albedo, primarily through the increase in the crystal size of the snow grains and through the accumulation of litter on the snow surface.

Dozier *et al.* (1981) have reported a study of how satellite data in the near-infrared region can be used to estimate snow grain size, at least qualitatively. Figure 4.5 shows how reflectance changes with snow grain size for two incidence angles. Although changes in albedo also occur in the visible region, they are not as great and are more affected by water equivalence and contamination by aerosols. Using NOAA-6 AVHRR data collected over Lake Winnipeg, Canada, the authors attributed the differences in measured albedo shown in Figure 4.6 to be primarily the result of increased grain size. It was also concluded that a satellite with a spectral response farther out into the infrared region (1.0–1.2 μm instead of 0.7–1.0 μm) would be more sensitive to grain size.

A hybrid approach to determine snow water content using digitally enhanced NOAA data and aircraft gamma-ray spectrometry has been used in Finland (Kuittinen 1986). Combining the satellite imagery with the gamma-ray data is a new approach for monitoring the relatively short melt season.

Snow water equivalence has also been estimated by a residual method in which measured runoff data are used to reconstitute the snow water content

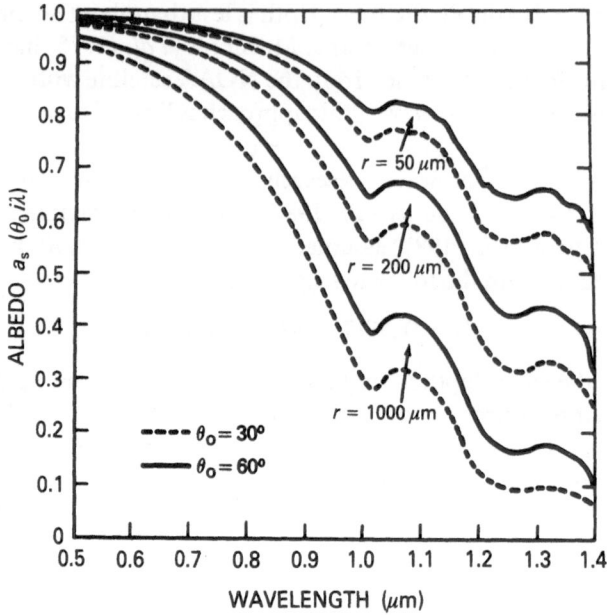

Figure 4.5 An illustration of the spectral direct beam albedo of semi-infinite snow for three grain radii and two solar zenith angles of 30° (dashed lines) and 60° (solid lines) (after Dozier *et al*. 1981).

(Martinec and Rango 1981). Landsat data have been used on a grid basis to track the disappearance of snow and, concurrently, the degree days necessary to melt the snow are calculated. These results can be used to improve the areal distribution of snow water equivalent in the basin and to correct winter precipitation measurements.

Many of the microwave measurements of snow properties have been conducted with truck and aircraft experiments. The results from a field data collection programme in which both active and passive microwave truck-mounted instruments were tested were reported by Stiles *et al*. (1981). Both the scattering coefficient and the emissivity were found to be sensitive to changes in snow water content and liquid water in the snow. Figure 4.7 shows the relationships for water equivalent and Figure 4.8 illustrates the diurnal response of both the active and passive systems to the formation of liquid water in the snow through melting, and its subsequent refreezing.

4.5 SNOWMELT RUNOFF

Predicting snowmelt runoff for either water supply or flood warning is a major task for hydrologists in many parts of the world. In some areas such as the

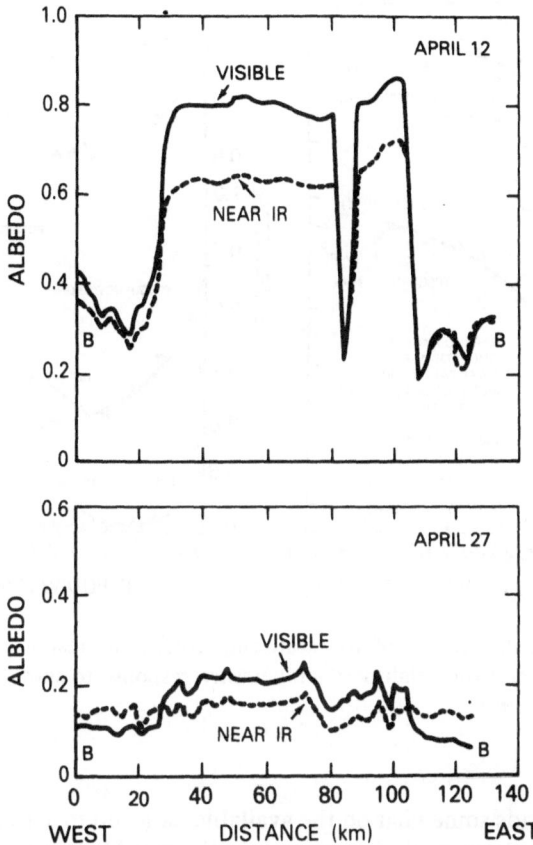

Figure 4.6 An illustration showing the trace of albedo (B–B) across southern Lake Winnipeg on April 12 (top) and April 27 (bottom). The low albedo in the bottom graph is attributed to much of the snow having melted (after Dozier *et al.* 1981), indicating that the snow is almost completely gone by the 27th (Dozier *et al.* 1981).

western United States; snowmelt runoff provides nearly all the water for industry, agriculture and domestic use. Accurate and timely prediction of snowmelt runoff is necessary for efficient reservoir management and planning the distribution of the water. Certain parts of the world are habitually plagued by flooding from rapidly melting snow. Prediction of peak flow rates and timing are necessary to provide adequate warning and emergency programmes.

General approach

Snowmelt runoff procedures using remote sensing have followed two distinct paths: an empirical approach and one based on modelling. The choice of

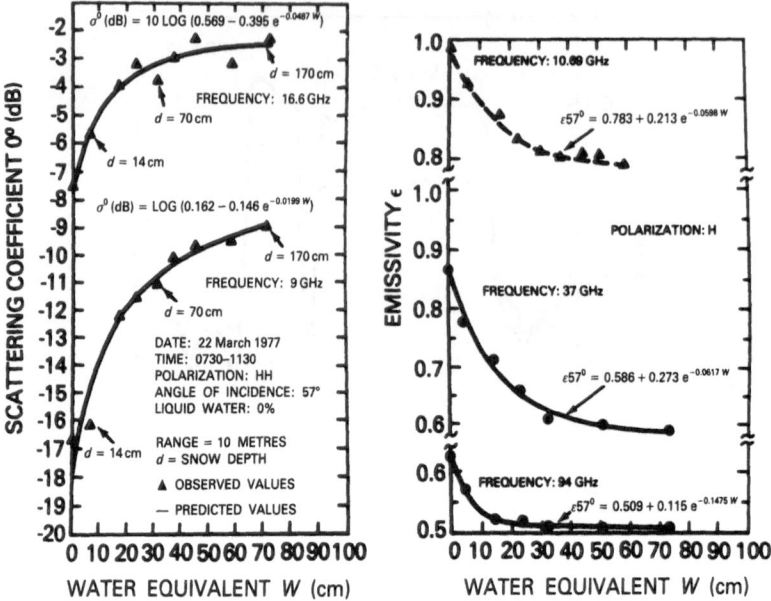

Figure 4.7 An illustration of the scattering coefficient response to snow water equivalent (left) and the brightness temperature response to snow water equivalent (after Stiles *et al*. 1981).

approach depends somewhat on the available data and to a great extent on the detail in output desired. Each approach is discussed below.

Empirical snowmelt prediction

Historically, data from snow courses have been used by hydrologists for their predictions with forecast models that have typically been of the multiple regression form

$$Y = a + bX1 + cX2 + dX3 + \dots \qquad (4.3)$$

where Y is the volume yield of runoff for the forecast period, and $X1$, $X2$, etc. are the snow water contents from each snow course. The coefficients a, b, etc. are developed from empirical data. Other variables such as autumn precipitation, base flow, etc. have also been included in specific models. Although useful and 'state of the art', these models have serious limitations. They are only valid for the basins for which they were developed and for the range of data used in the regression equations. In addition, snow course data are point measurements which at best serve only as indices of the snow in a

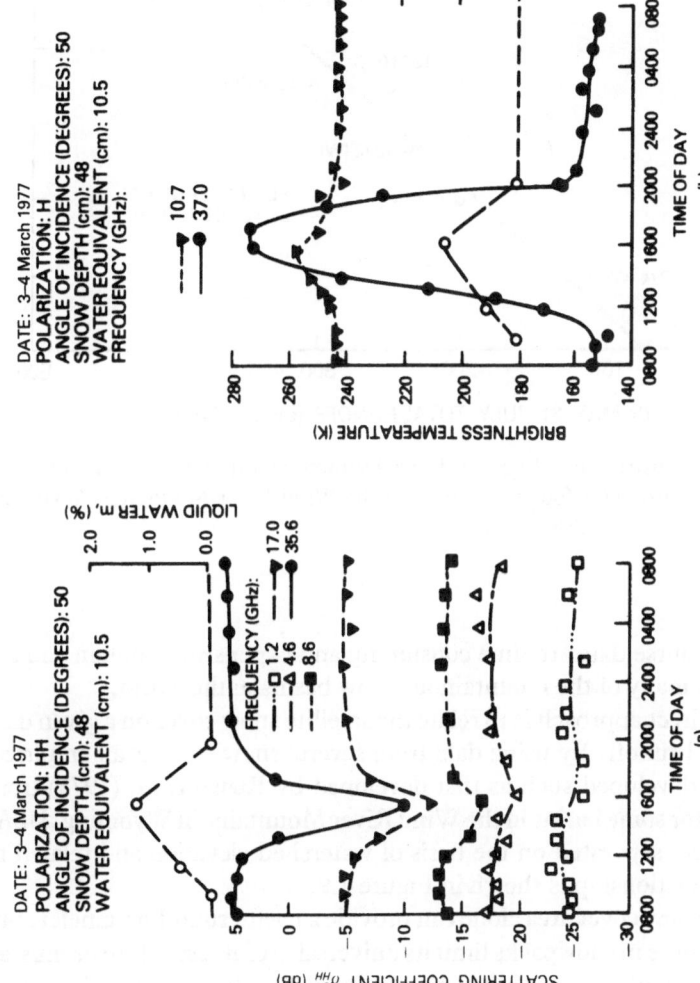

Figure 4.8 An illustration of the sensitivity of the scattering coefficient (left) and brightness temperature (right) to the presence of liquid water in the snowpack (after Stiles *et al.* 1981).

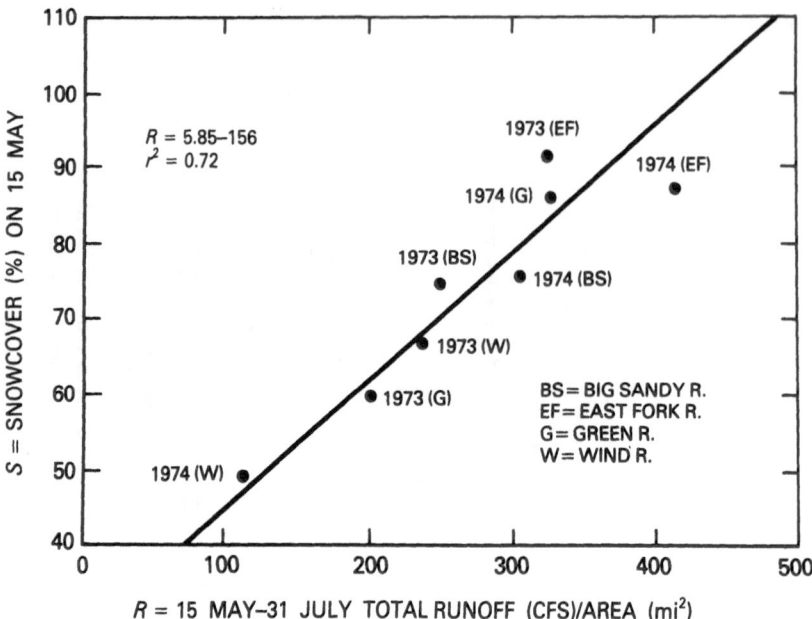

Figure 4.9 An illustration of Landsat-derived snowcover estimates against measured runoff (1973 and 1974) for four watersheds in the Wind River Mountains, Wyoming (after Rango *et al.* 1975). CFS = cubic feet per second.

basin. Snow course data are time consuming and expensive to obtain and are unavailable in many of the mountainous snow basins in the world.

The most direct approach is to relate the satellite snow cover on a given date to the seasonal runoff. By using data from several snow seasons an empirical curve can be developed such as that developed by Rango *et al.* (1975) from Landsat data for some basins in the Wind River Mountains in Wyoming, USA. Snow cover was separated on the basis of watershed elevation and related to runoff. This relationship is shown in Figure 4.9.

The use of snow cover area alone can provide a useful runoff parameter, but annual differences in snowpacks limit its universal usefulness. More promising results have been obtained using a concept known as the dimensionless basin snow cover depletion–accumulation curves. Use of dimensionless snow cover area (snow cover area on a given date normalized by the total basin area) and dimensionless accumulated runoff (accumulated runoff at the date normalized by the total seasonal runoff) effectively compensates for the season-to-season variations in snowpack depth and water content and weather. Figure 4.10 shows a dimensionless depletion–accumulation curve for the Conejos River in Colorado.

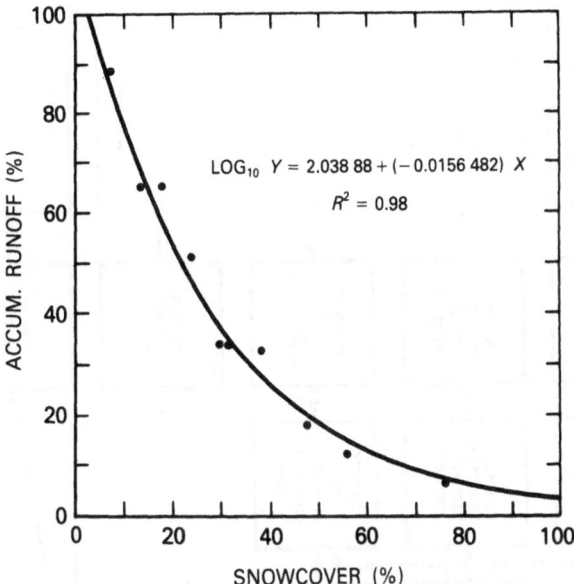

Figure 4.10 An illustration of a dimensionless depletion–accumulation curve showing composite snowcover plotted versus accumulated runoff (after Thompson 1975).

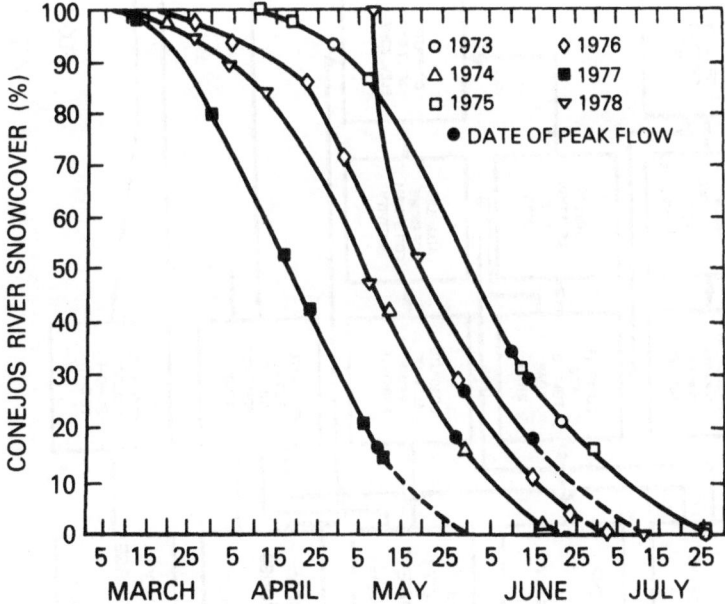

Figure 4.11 An illustration of snowcover depletion curves for the Conejos River (after Thompson 1975).

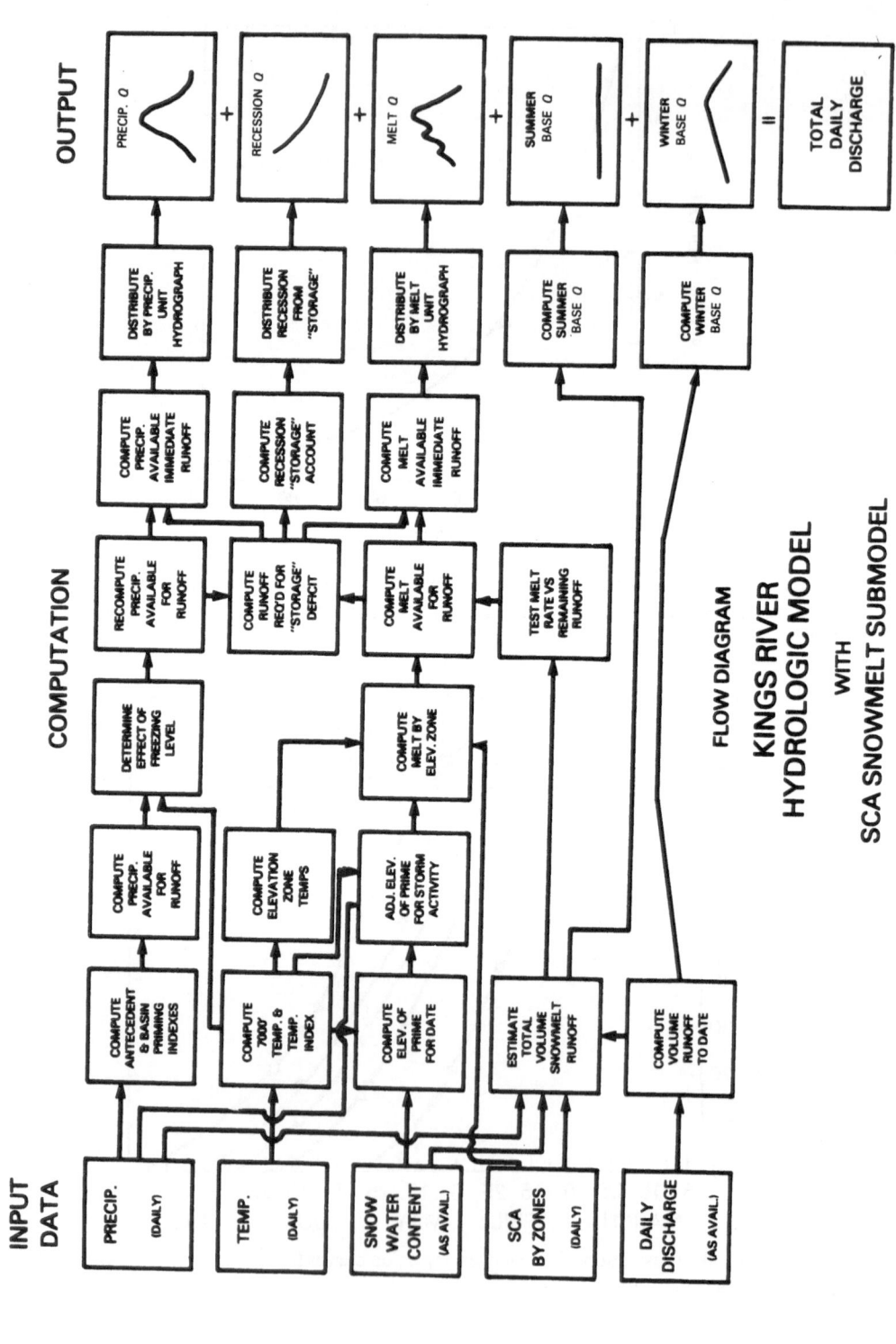

FLOW DIAGRAM
KINGS RIVER
HYDROLOGIC MODEL
WITH
SCA SNOWMELT SUBMODEL

Snow cover depletion curves are another empirical approach that use satellite-derived snow cover area. The snow cover depletion curves show a melt sequence that is repeatable from one year to the next (Hall and Martinec 1985). The displacement of the curves from one year to the next is related to the snowpack water content. That is, in low snowpack years the melt begins earlier and the runoff is reduced, and vice versa. Although these curves are developed for one basin, they may possibly be transposed to other nearby basins as an estimate of the melt process in the nearby basin. Figure 4.11 shows the family of snow cover depletion curves developed for the Conejos River basin in Colorado.

Modelling snowmelt

A number of models for predicting and simulating snowmelt runoff have been developed or have been modified from existing models to incorporate satellite-derived snow cover data.

The California Department of Water Resources has taken an existing hydrologic model for the Kings River basin, removed the original snowmelt component, and replaced it with a procedure based on the snow cover area (SCA). A block diagram of this model is shown in Figure 4.12. The model is based on the premise that snowmelt does not occur until the snow has been primed. Experience has shown that this is an elevation-dependent phenomenon. Thus, the basin is delineated into specific elevation zones known as the 'elevation of prime'. The basic empirical relationship for computing snowmelt at a given elevation of prime in this basin is

$$E_p = 3.1 \times K \times (1.009)^D + 0.017 \times K \times (T1 - 100/K) \qquad (4.4)$$

where E_p is the elevation of prime (100 ft (30.5 m)) which is the maximum elevation at which snowmelt can occur, D is the number of days since 1 February, $T1$ is the decayed annual temperature taken as the degree days at 7000 ft (2135 m) since 1 January and using a decay factor of 0.96, and K is a variable that affects the elevation of prime that is related to the snow water content.

A fairly complex subalpine model has been developed by the US Forest Service to simulate daily streamflow (Leaf and Brink 1973). The model schematic is shown in Figure 4.13. The model accounts for winter snow accumulation, a short-wave and long-wave radiation balance, snowpack condition and the resulting snowmelt. When applying the model, the basin is

Figure 4.12 A schematic of the flow diagram for the Kings River hydrologic model and the SCA snowmelt submodel (after Hannaford and Hall 1980).

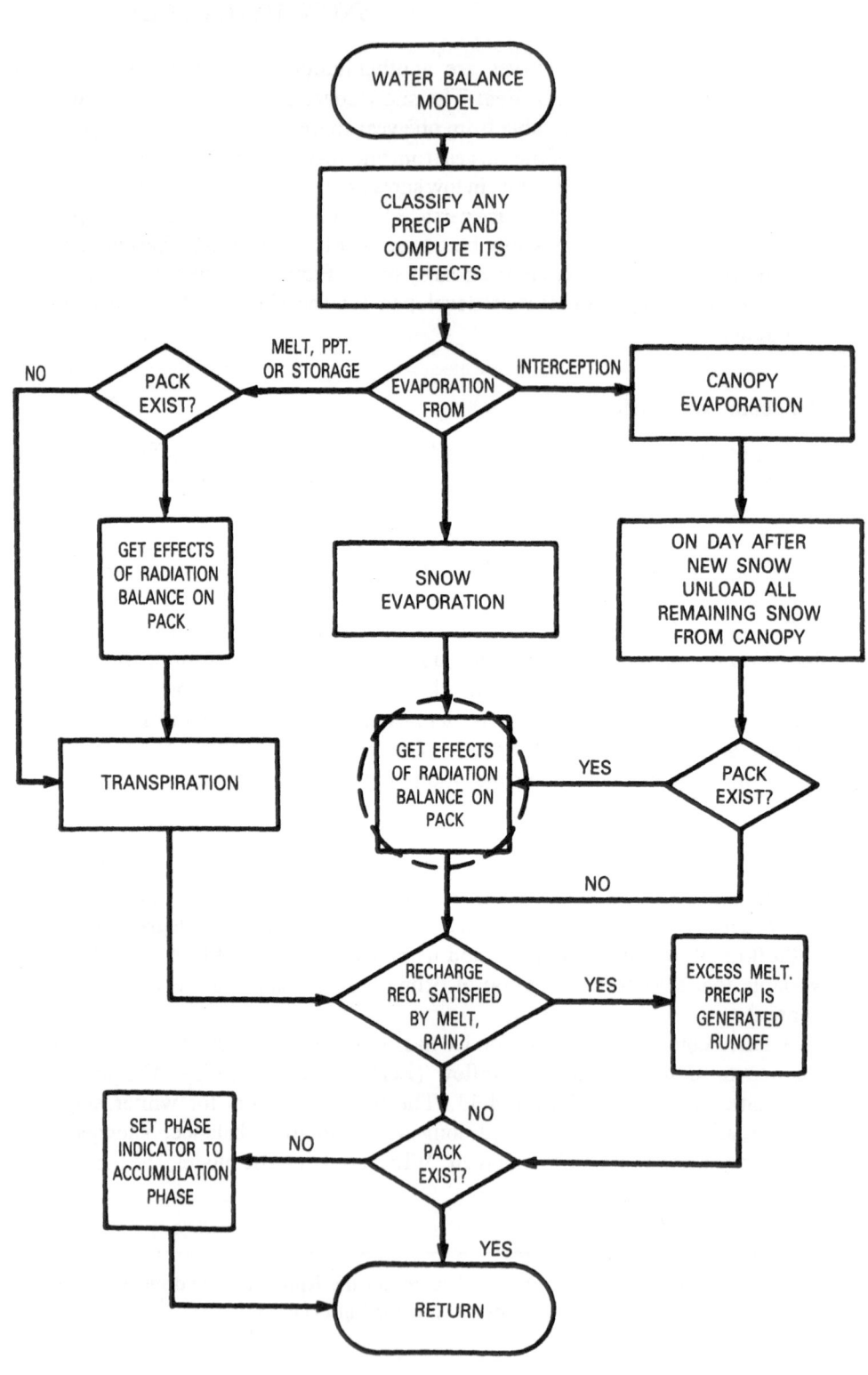

divided into as many as 25 subunits of relatively uniform aspect, slope and cover. During the snowmelt season the model is updated by real-time snow telemetry data and satellite snow cover area data.

In the Pacific north-western United States, satellite snow cover data are being used operationally in the streamflow synthesis and reservoir regulation (SSARR) model. After initializing the model, the model parameters (snow cover area, soil moisture, initial melt rate, base flow and seasonal volume) are adjusted until the forecast and measured hydrographs agree within a predetermined tolerance. The snow cover area data are very important, especially during the initial adjustment phases. In test cases for five basins over a 6 year period, the addition of satellite snow cover data resulted in definite but statistically insignificant improvements (Dillard and Orwig 1979).

The snowmelt runoff model (SRM) for simulating snowmelt is one of the better known and most thoroughly tested empirical models available. SRM has been developed to simulate and forecast daily streamflow in mountain basins where snowmelt is a major component of the annual water balance. SRM is a degree day model that uses the percentage of the basin or elevation zone covered by snow as the primary input. SRM was developed by Martinec (1975) for small European basins. With the advent of satellite snow cover data in the 1970s, the model was shown to be usable in larger basins. Using Landsat data, SRM has been successfully run for various-sized basins in Europe and the United States (Rango and Martinec 1979; Rango 1980; Rango 1983; Martinec and Rango 1986). Kawata et al. (1988) have shown how Landsat MSS data can be used with the SRM to predict runoff and manage water levels in the Sai River dam, near Kanazawa, Japan. The model is a good compromise between the amounts and types of data usually available.

Applications of SRM usually involve simulating snowmelt runoff for basins with some hydrologic data, specifically measured discharge and one or more meteorological stations that measure precipitation and temperature. The minimum required inputs to the model consist of periodic snow cover area and daily temperature and precipitation. Each day during the snowmelt season, the water produced from snowmelt is computed and superimposed on the base flow to yield the total basin discharge according to the equation

$$Q_{n+1} = c_n \left[a_n (T_n + \Delta T_n) S_n + P_n \right]$$
$$\times A(0.01/86400)(1 - k_{n+1}) + Q_n k_{n+1} \qquad (4.5)$$

where Q is the average daily discharge ($m^3 s^{-1}$), c the runoff coefficient expressing the losses as a ratio (runoff/precipitation), a the degree day factor

---

Figure 4.13   A schematic of the flow diagram for the subalpine water balance model (after Leaf and Brink 1973).

$(\text{cm}\,{}^\circ\text{C}^{-1}\,\text{day}^{-1})$ indicating the snowmelt depth resulting from 1 degree day, T the number of degree days $({}^\circ\text{C}\,\text{day})$, ΔT the adjustment by temperature lapse rate necessary because of the altitude difference between the temperature station and the average hypsometric elevation of the basin or zone, S the ratio of the snow-covered area to the total area, P the precipitation contributing to runoff (cm) (a preselected threshold temperature, T_{crit}, determines whether this contribution is rainfall and immediate), A the area of the basin or zone (m^2), $(0.01/86400)$ the conversion from cm $\text{m}^2\,\text{day}^{-1}$ to $\text{m}^3\,\text{s}^{-1}$, k the recession coefficient indicating the decline of discharge in a period without snowmelt or rainfall, $k = (Q_{m+1}/Q_m)$ $(m, m+1)$ are the sequence of days during a true recession flow period), and n the sequence of days during the discharge computation period. Equation (4.5) is written for a time lag between the daily temperature cycle and the resulting discharge cycle of 18 h. As a result, the number of degree days measured on the nth day corresponds to the discharge on the $(n+1)$th day. Different lag times will result in the proportioning of day n snowmelt between discharges occurring on days n, $n+1$ and possibly $n+2$.

There are a number of steps a user must take before the model can be run. The user must determine the physical characteristics of the basin and select the model variables and parameters. After the basin boundary has been defined by the stream gauge site, the basin must be subdivided into elevation zones separated by about 500 m. The elevation zones are created in recognition that snowmelt is very elevation dependent and thus the model is applied to each of the zones to distribute the rate of snowmelt spatially. A hypsometric curve is developed for the basin and the mean elevation for each zone is determined graphically.

Input data to the model are air temperature, precipitation and snow cover area. Ideally, temperature and precipitation would be measured within the basin and at each mean hypsometric elevation for each zone. Seldom is this the case in real basins. Usually these data must be extrapolated from one or more stations, some of which may not be in the basin.

Snow cover data are used to construct snow cover depletion curves for each zone of the basin. Usually the snow cover data are planimetered from remote sensing satellite data.

Model parameters include a runoff coefficient, a degree day factor, a recession coefficient and a time lag. Each of these can be chosen and modified by the user to adjust the simulation results. The model is set up with default values and the user's manual (Martinec et al. 1983) describes how the choices should be made.

Typical applications usually require several iterations that involve changes in parameters or the subdivision of the basin into equal elevation zones. Depending on how good the initial simulations were and what type of discrepancies exist between the measured and simulated hydrographs, the user

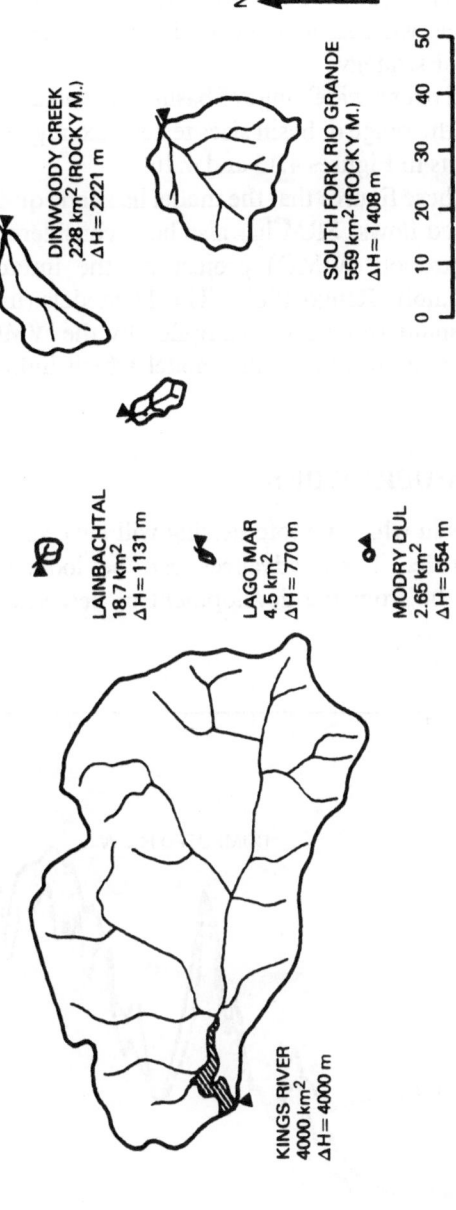

Figure 4.14 An illustration of the range of areas and total basin relief (ΔH) of basins in which SRM has been applied (after Rango 1983).

chooses different strategies to improve the simulations. In a basin where the model has been applied before, the user works with a sequential plan based on accumulated knowledge, whereas in a new basin or area the user must resort more to a trial-and-error strategy.

SRM has been tested over a wide range of basin sizes and areas of the world. Figure 4.14 illustrates the range of basin sizes tested. Examples of simulations are shown for two basins in Figures 4.15 and 4.16.

It can be seen from these figures that the model has done quite a good job of simulating the measured flows. SRM has also been considered in the World Meteorological Organization (WMO) project on the intercomparison of models of snowmelt runoff (Rango 1985). The 11 models (including SRM) were tested using six standard data sets compiled by the WMO (1982). The results confirmed previous testing of the model which indicated universal usage.

4.6 FUTURE CONSIDERATIONS

There are several areas in which remote sensing will have an impact on snow hydrology in the future. These advances can be looked at from two perspectives: one will be from the development of new sensors, especially

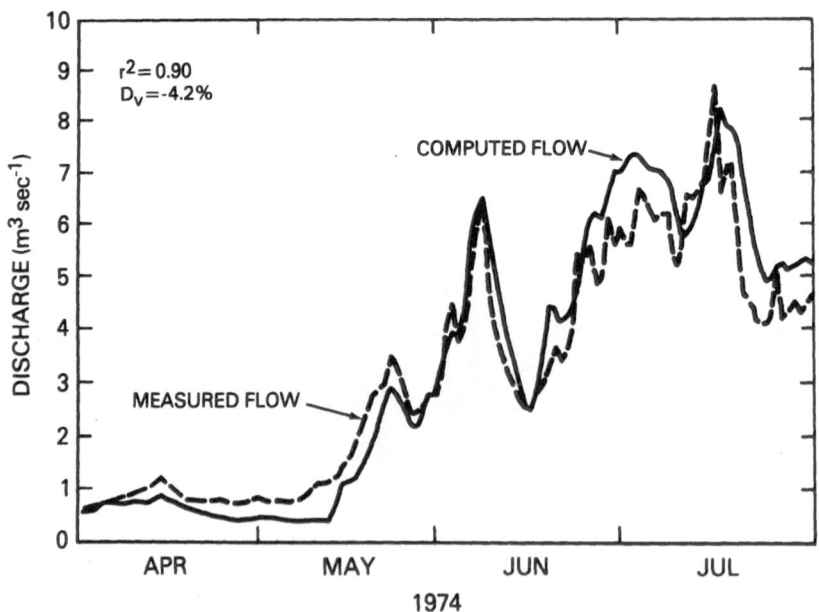

Figure 4.15 An illustration of discharge simulation for the Dischma basin (43.3 km²), Switzerland, using SRM (after Rango 1985).

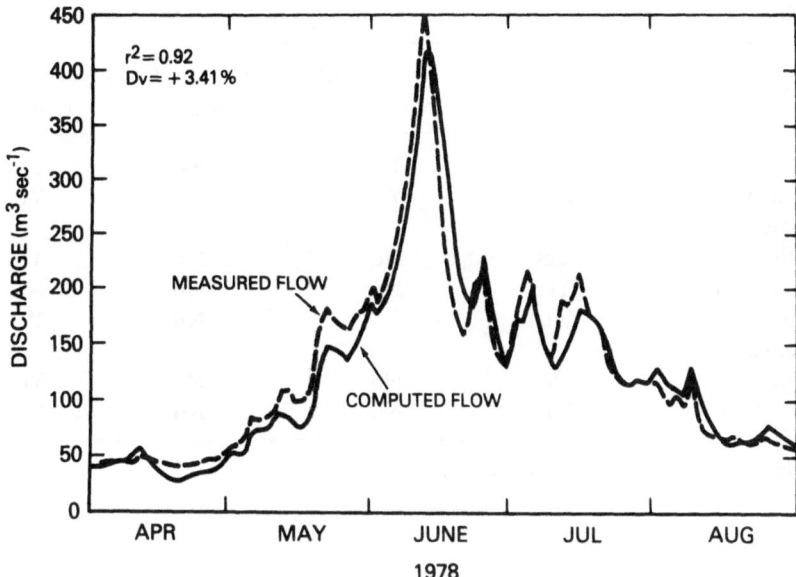

Figure 4.16 An example of discharge simulation for the Durance River basin (2170 km^2), France, using SRM (after Rango 1985).

microwave sensors; and the second will be the modification of models and the utilization of new types of data and databases.

New sensors and methods of analysis will have a major impact on remote sensing of snow and snowmelt hydrology. The snow properties which are discernible from various regions of the spectrum are summarized in Table 4.3. It can be seen that in most cases simultaneous measurements would need to be made at several wavelengths to determine specific snow properties. The combination of visible, near-infrared and microwave measurements at various spatial scales as planned for the Eos mission will provide much of the necessary data for sophisticated snow measurements and analysis of the snowmelt process.

New instruments, together with innovative methods of analysis, should provide new insight to the solution of snow hydrology problems. For example, Danes and Danes (1988) demonstrated how the difference in polarization between the SMMR 37 GHz and 19 GHz passive microwave measurements could be related to snow water equivalent. Their research showed good results for areas of sparse vegetation in the upper Colorado River basin.

New models that are developed to use remote sensing data will also improve the snow hydrology predictions. SRM is being modified for forecasting (Rango and van Katwijk 1990). To do so, a series of snow depletion curves must be

Table 4.3   Properties affecting albedo and emissivity of snow (after Warren 1982)

	Visible reflectance	Near infrared reflectance	Thermal infrared emissivity	Microwave emissivity
Grain size	*	Yes	No	Yes
Zenith (or nadir) angle	No	Yes	Yes	Yes
Depth	Yes	No	No	Yes
Contaminants	Yes	No	No	No
Liquid water content	No	*	No	Yes
Density	No	No	No	Yes
Temperature	No	No	No	Yes

* Only for thin snowpack or if impurities present.

developed (Hall and Martinec 1985). These snow depletion curves can usually be developed from historic Landsat, SPOT or AVHRR data.

Another model that is perhaps more physically based than SRM has been recently modified to use remote sensing data of the snowpack. The PRMS (precipitation–runoff modelling system) (Leavesley *et al.* 1983) is a modular, distributed parameter watershed model which treats the snowpack as a two-layer system. Heat and mass (rain or snow) are transferred across a 3–5 cm surface layer to the main body of the snowpack, which is maintained and modified both as a water reservoir and a heat reservoir.

The merging of remote sensing data with digital elevation modelling (DEM) and geographical information systems (GIS) enables different types of data to be combined objectively and systematically. DEM is used to normalize imagery by using the elevation of the sun and the slope, aspect and elevation of the terrain (Baumgartner 1988; Miller *et al.* 1982). GIS are helpful for combining vegetation masks with satellite imagery (Keller 1987).

Snow hydrology was one of the first areas in water resources to make effective use of remote sensing data, and it appears to be able to take advantage of future developments as they become available. The parallel development of understanding sensor response and modelling is a major factor in the present state of the field.

REFERENCES

Baumgartner, M. F. (1988) Snowmelt runoff simulation based on snow cover mapping using digital Landsat-MSS and NOAA/AVHRR data. *USDA-ARS, Hydrology Lab. Tech. Rep.*

Bowley, C. J., Barnes, J. C. and Rango, A. (1981) Applications systems verification and transfer project, vol. VIII: Satellite snow mapping and runoff prediction handbook. *NASA Tech. Pap.* 1829, Goddard Space Flight Center, Greenbelt, MD.

Brown, A. J., Hannaford, J. F. and Hall, R. L. (1979) Application of snow covered area to runoff forecasting in selected basins of the Sierra, Nevada, California. *Proc. Final Workshop on Operational Applications of Satellite Snow Cover Observations*, NASA CP-2116, pp. 185–200.

Burke, H. K., Bowleg, C. J. and Barnes, J. C. (1984) Determination of snowpack properties from satellite passive microwave measurements. *Remote Sensing Environ.* 15, 1–20.

Carroll, S. S. and Carroll, T. R. (1989) Effect of forest biomass on airborne snow water equivalent estimates obtained by measuring terrestrial gamma radiation. *Remote Sensing Environ.* 7, 313–20.

Carroll, T. R. and Allen, M. (1988) Airborne gamma radiation snow water equivalent and soil moisture measurements and satellite areal extent of snow cover measurements. *A User's Guide*, Version 3.0, National Weather Service, NOAA, Minneapolis, MN.

Carroll, T. R. and Vadnais, K. G. (1980) Operational airborne measurement of snow water equivalent using natural terrestrial gamma radiation. *Proc. 48th Annu. Western Snow Conf., Laramie, WY*, Western Snow Conference, pp. 97–106.

Carroll, T. R. and Vose, G. D. (1984) Airborne snow water equivalent measurements over a forested environment using terrestrial gamma radiation. *Proc. 41st Annu. Eastern Snow Conf., New Carrollton, MD.*, Eastern Snow Conference, p. 19.

Castruccio, P. A., Loats, Jr, H. L., Lloyd, D. and Newman, P. B. (1980) Cost/benefit analysis for the operational applications of satellite snow cover observations (OASSO). *Proc. Final Workshop on Operational Applications of Satellite Snow Cover Observations*, NASA CP-2116, pp. 239–54.

Chang, A., Foster, J. and Hall, D. K. (1987) NIMBUS-7 derived global snow cover parameters. *Ann. Glaciol.* 9, 39–44.

Crane, R. G. and Anderson, M. R. (1984) Satellite discrimination of snow/cloud surfaces. *Int. J. Remote Sensing* 5, 213–23

Danes, Z. F. and Danes, P. L. R. (1988) Polarization of passive microwave signals as indicator of snow water equivalent. *Proc. IGARSS'88 Symp., Edinburgh, Scotland*, ESA SP-284, pp. 441–2.

Dey, B., Goswami, D. C. and Rango, A. (1983) Utilization of satellite snow-cover observations for seasonal streamflow estimates in the western Himalayas. *Nord. Hydrol.* 257–266.

Dillard, J. P. and Orwig, C. E. (1979) Use of satellite data in runoff forecasting in the heavily forested, cloud covered Pacific northwest. *Proc. Final Workshop on Operational Applications of Satellite Snow Cover Observations*, NASA CP-2116, pp. 127–150.

Dozier, J. (1984) Snow reflectance from Landsat-4 thematic mapper. *IEEE Trans. Geosci. Remote Sensing* **GE-22**, 323–8.

Dozier, J. and Marks, D. (1987) Snow mapping and classification from Landsat Thematic Mapper. *Ann. Glaciol.* 9, 97–103.

Dozier, J., Schneider, S. R. and McGinnis, Jr, D. F. (1981) Effect of grain size and snowpack water equivalence on visible and near-infrared satellite observations of snow. *Water Resour. Res.* 17, 1213–21.

Foster, J. L., Hall, D. K. and Chang, A. T. C. (1987) Remote sensing of snow. *Eos* **68**(32), 681–4.

Foster, J. L., Hall, D. K., Chang, A. T. C. and Rango, A. (1984) An overview of passive microwave snow research and results. *Rev. Geophys. Space Phys.* **22**, 195–208.

Frank, C., Itten, K. I. and Staenz, K. (1988) Improvement in NOAA-AVHRR snowcover determination for runoff prediction. *Proc. IGARSS'88 Symp., Edinburgh, Scotland*, ESA SP-284, pp. 433–5.

Glynn, J. E., Carroll, T. R., Holman, P. B. and Grasty, R. L. (1988) An airborne gamma ray snow survey of a forest covered area with a deep snowpack. *Remote Sensing Environ.* **26**, 149–60.

Griggs, M. (1968) Emissivities of natural surfaces in the 8- to 14-micron spectral region. *J. Geophys. Res.* **73**, 7545–51.

Hall, D. K. and Martinec, J. (1985) *Remote Sensing of Ice and Snow*, Chapman and Hall, London.

Hallikainen, M. (1984) Retrieval of snow water equivalent from NIMBUS-7 SMMR data: Effect of land cover categories and weather conditions. *IEEE J. Ocean Eng.* **OE-9**, 372–6.

Hannaford, J. F. and Hall, R. L. (1980) Application of satellite imagery to hydrologic modeling snowmelt runoff in the southern Sierra Nevada. *Proc. Final Workshop on Operational Applications of Satellite Snow Cover Observations*, NASA CP-2116, pp. 201–22.

Harrison, A. R. and Lucas, R. M. (1989) Multispectral classification of snow using NOAA AVHRR imagery. *Int. J. Remote Sensing* **10** (4 and 5), 907–16.

Kawata, Y., Kusaka, T. and Veno, S. (1988) Snowmelt runoff estimation using snowcover extent data and its application to optimum control of dam water level. *Proc. IGARSS'88 Symp. Edinburgh, Scotland*, ESA SP-284, pp. 439–40.

Keller, M. (1987) Ausaperungskartierung mit Landsat-MSS Daten zur Erfassung oekologische Einflussgroessen im Gebirge. *Remote Sensing Series*, No. 10, PhD Thesis, University of Zürich.

Kuittinen, R. (1986) Determination of areal snow water equivalent values using satellite imagery and aircraft gamma ray spectrometry. *Hydrologic Applications of Space Technology (Proc. Cocoa Beach Workshop, Florida)*, IAHS Publ. No. 160, pp. 181–9.

Kunzi, K. F., Patil, S. and Rott, H. (1982) Snow-covered parameters retrieved from NIMBUS-7 SMMR data. *IEEE Trans. Geosci. Remote Sensing* **GE-20**, 452–67.

Leaf, C. F. (1969) Aerial photographs for operational streamflow forecasting in the Colorado Rockies. *Proc. 37th Western Snow Conf., Salt Lake City, UT.*

Leaf, C. F. and Brink, G. E. (1973) Hydrologic simulation model of Colorado subalpine forest. *USDA Forest Service, Res. Pap.* RM-107, Fort Collins, CO.

Leavesley, G. H., Lumb, A. M. and Saindon, L. G. (1987) A micro-computer-based watershed-modeling and data-management system. *Proc. 55th Annu. Western Snow Conf., Vancouver, BC*, Western Snow Conference, pp. 108–17.

Lucas, R. M., Harrison, A. W. and Barrett, E. C. (1989) A multispectral snow area algorithm for operational 7-day snow cover monitoring. *Remote Sensing and Large Scale Global Processes, Proc. IAHS 3rd Int. Assembly, Baltimore, MD, IAHS Publ. No. 1986*, pp. 161–6.

McGinnis, D. F., Pritchard, J. A. and Wiesnet, D. R. (1975) Determination of snow depth and snow extent from NOAA-2 satellite very high resolution radiometer data. *Water Resour. Res.* **11**, 892–902.

Martinec, J. (1975) Snowmelt-runoff model for streamflow forecasts. *Nord. Hydrol.* **6**, 145–54.

Martinec, J. and Rango, A. (1981) Areal distribution of snow water equivalent evaluated by snow cover monitoring. *Water Resour. Res.* **17**, 1480–8.

Martinec, J., Rango, A. and Major, E. (1983) *The Snowmelt-Runoff Model (SRM) User's Manual.* NASA Ref. Publ. 1100, Washington, DC.

Martinec, J. and Rango, A. (1986) Parameter values for snowmelt runoff modeling. *J. Hydrol.* **84**, 197–219.

Matson, M. and Wiesnet, D. R. (1981) New data base for climate studies. *Nature* **289** 451–56.

Meier, M. J. and Evans, W. E. (1975) Comparison of different methods for estimating snow cover in forested, mountainous basins using LANDSAT (ERTS) images. *Operational Applications of Satellite Snow Cover Observations,* NASA SP-391, Washington DC, pp. 215–34.

Miller, W.A., Shasby, M. B., Rhode, W. G. and Johnson, G. R. (1982) Developing in-place data bases by incorporating digital terrain data into the Landsat classification process. *Place Resource Inventories: principles and practices, Proc. Workshop, 9–14 August, 1981, University of Maine, Orono,* Sponsored by the American Society of Photogrammetry.

Odegaard, H. A., Andersen, T. and Ostrem, G. (1979) Application of satellite data for snow mapping in Norway. *Proc. Final Workshop on Operational Applications of Satellite Snow cover Observations,* NASA CP-2116, pp. 93–106.

Ostrem, G., Andersen, T. and Odegaard, H. (1981) Operational use of satellite data for snow inventory and runoff forecast. *Satellite Hydrology,* American Water Resources Association, Minneapolis, MN, pp. 230–4.

Patil, S. Kunzi, K. F. and Rott, H., (1981) The global snow cover seen by the NIMBUS-7 Scanning Multichannel Microwave Radiometer (SMMR). *Proc. 11th Eur. Microwave Conf., Amsterdam,* Microwave Exhibitions and Publications, Amsterdam, The Netherlands, pp. 227–32.

Rango, A. (1980) Operational applications of satellite snow cover observations. *Water Res. Bull.* **16**, 1066–73.

Rango, A. (1983) Application of a simple snowmelt-runoff model to large river basins. *Proc. 51st Western Snow Conf.* Western Snow Conference, pp. 89–99.

Rango, A. (1985) Results of the snowmelt-runoff model in an international test. *Proc. 53rd Western Snow Conf., Boulder, Co.,* Western Snow Conference, pp. 99–106.

Rango, A., Chang, A. T. C. and Foster, J. L. (1979) The utilization of spaceborne microwave radiometers for monitoring snowpack properties. *Nord. Hydrol.* **10**, 25–40.

Rango, A. and Itten, K. (1976) Satellite potentials in snowcover monitoring and runoff prediction. *Nord. Hydrol.* **7**, 209–30.

Rango, A. and Martinec, J. (1979) Application of a snowmelt-runoff model using Landsat data. *Nord. Hydrol.* **10**, 225–38.

Rango, A., Martinec, J., Foster, J. and Marks, D. (1983) Resolution in operational remote sensing of snow cover. *Hydrological Applications of Remote Sensing and Remote*

Data Transmission, Proc. Hamburg Symp., August, 1983, IAHS Publ. No. 145, pp. 371–82.

Rango, A., Salomonson, V. V. and Foster, J. L. (1975) Employment of satellite snow cover observations for improving seasonal runoff estimates. *Operational Applications of Satellite Snow Cover Observations*, NASA SP-391, Washington, DC, pp. 157–74.

Rango, A., Salomonson, V. V. and Foster, J. L. (1977) Seasonal streamflow estimation in the Himalayan region employing meteorological satellite snow cover observations. *Water Resour. Res.* 13, 109–12.

Rango, A. and van Katwijk, V. (1990) Development and Testing of a Snowmelt-runoff Forecasting Technique. *Water Resvour. Bull.* 26(1), 135–44.

Shafer, B. A. and Leaf, C. F. (1979) Landsat derived snowcover as an input variable for snowmelt runoff forecasting in Central Colorado. *Proc. Final Workshop on Operational Applications of Satellite Snow Cover Observations*, NASA CP-2116, pp. 151–69.

Shcheglova, O. P. and Chernov, V. Yu. (1982) Seasonal snow line within the Fergana Basin and possibilities for using it in hydrological forecasting. In *Izucheniye Gidrologicheskogo Tsikla Aerokomicheskimi Metodami* (eds K. Ya. Kondratyev and Yu V. Kurilova) Academy of Sciences of the USSR, Soviet Geophysical Committee, Moscow, pp. 36–41.

Stiles, W. H., Ulaby, F. T. and Rango, A. (1981) Microwave measurements of snowpack properties. *Nord. Hydrol.* 12, 143–66.

Thompson, A. G. (1975) Utilization of Landsat monitoring capabilities for snow cover depletion analysis. *Operational Applications of Satellite Snow Cover Observations*, NASA, SP-391, Washington, DC, pp. 113–27.

Vershinina, L. K. (1985) The use of aerial gamma surveys of snowpack for spring snowmelt runoff forecasts. *Hydrological Applications of Remote Sensing and Remote Data Transmission, Proc. Hamburg Symp, August 1983, IAHS Publ. No.* 145, pp. 411–20.

Warren, S. G. (1982) Optical properties of snow. *Rev. Geophys. Space Phys.* 20, 67–89.

Wiesnet, D. R. and Matson, M. (1975) Monthly winter snowline variation in the Northern Hemisphere from satellite records. *NOAA Tech. Memo. NESS* 74, National Environmental Satellite Service, Washington, DC.

World Meteorological Organization (1982) WMO project for the intercomparison of conceptual models of snowmelt runoff. *Hydrological Aspects of Alpine and High Mountain Areas, Proc. Exeter Symp., IAHS Publ. No.* 138, pp. 193–202.

5

Evapotranspiration

5.1 INTRODUCTION

Evapotranspiration is the loss of water from the Earth's surface in vapour form. It occurs as evaporation from open water and moist soil surfaces and as transpiration from living plants as part of their respiration and photosynthetic processes. As far as the local or regional water balance is concerned, evapotranspiration (ET) is a significant and often the dominant water flux leaving the Earth's land surface. In arid regions, nearly all the input in the form of rain is lost through ET. Even in humid regions, one-half or more of the water balance can be attributed to ET.

Evaporation and transpiration estimates are required for several purposes, including irrigation scheduling, reservoir losses, water balance calculations, runoff prediction, and meteorological and climatological studies. Long-term estimates of ET may be made using water balance methods, as in lysimetry or evaporation pans, or at a larger scale in river basins. However, there is considerable uncertainty in these measurements, particularly for shorter time periods, and it is often difficult to generalize the measurements to large areas. Furthermore, many areas of interest lack the necessary field instrumentation to make even the most general ET estimates. For such areas, Thornthwaite (1948) suggested an empirical approach to estimate long-term evaporation based on routine meteorological observations, principally mean monthly temperature. Penman (1948) derived a more physically based expression which uses standard meteorological data to estimate potential ET, namely that which would occur from a short grass cover with unlimited water supply. The Jensen–Haise method (Jensen and Haise 1963) is another empirical method that uses climatological and solar radiation data to estimate ET. Although this approach has served as the theoretical basis for most of the currently used models, their application is still limited to locations where the standard meteorological data are available. In principle, remotely sensed measurements offer methods for extending these models to much larger areas, including those areas where data may be sparse. Although no method is as yet operational, the utility of such measurements is sufficiently great to encourage further work.

5.2 GENERAL APPROACH

To understand better how remote sensing may contribute, it is instructive to examine first the basic energy and moisture balance equations.

The energy balance of a soil or vegetated surface is governed by a conservation equation. In the absence of precipitation or advection, we may write

$$R_n + G + H + LE = 0 \qquad (5.1)$$

where R_n is the net radiation flux, G the soil heat flux, H the sensible heat flux and LE the latent heat flux in the atmospheric boundary layer. The latter is the product of the water vapour flux, E, and the latent heat of vaporization of water per unit mass, L.

The net radiation flux is the sum of the incoming and outgoing short- and long-wave radiation fluxes:

$$R_n = (1 - \alpha_s) R_s + R_L - e_L \sigma T_s^4 \qquad (5.2)$$

where R_s is the incoming short-wave radiation and R_L the incoming long-wave radiation, α_s is the short-wave soil/crop reflectivity or albedo, e_L is the long-wave emissivity, σ is the Stefan–Boltzmann constant and T_s is the surface temperature in degrees kelvin.

The soil heat flux is related to the temperature gradient in the soil dT/dz and to the thermal conductivity, λ:

$$G = \lambda \, dT/dz \qquad (5.3)$$

This relationship leads to a form of the diffusion equation:

$$d(\lambda \, dT/dz)/dz = C_v \, dT/dt \qquad (5.4)$$

where C_v is the volumetric heat capacity of the soil. This equation may be solved numerically to give a temperature profile in the soil, taking either T_s or G as the upper boundary condition and some known temperature at a given depth in the soil as the lower boundary condition.

The sensible and latent heat fluxes may both be estimated from the bulk transport equations. The sensible heat flux H may be expressed as (Brown and

Rosenberg 1973)

$$H = \varrho C_p (T_a - T_s)/r_a \tag{5.5}$$

where ϱ is the air density, C_p the specific heat of the air and r_a the aerodynamic resistance. The latent heat flux may be expressed as (Monteith 1973)

$$LE = \frac{\varrho c_p (e_a - e_s)}{\gamma (r_a + r_s)} \tag{5.6}$$

where γ is the psychometric constant, e_a the atmospheric vapour pressure in the boundary layer, e_s the saturated vapour pressure at the temperature T_s, and r_s the stomatal diffusion resistance to water vapour transport.

Equations (5.2), (5.3), (5.5) and (5.6) are a mathematical model of the surface energy balance in terms of the soil surface temperature and a set of meteorological measurements, which at least in principle are measured or estimated routinely. The energy and moisture fluxes are intimately related because the latent heat flux gives the amount of water evaporated and transpired.

More complete discussions of the ET process are given in the review paper by Monteith (1981) and the book by Brutsaert (1982).

5.3 CURRENT APPLICATIONS

In recent years, there has been much progress in the remote sensing of a number of parameters which can contribute to the estimation of ET. These include surface temperature, surface soil moisture, surface albedo, vegetative cover and incoming solar radiation. There has been little progress in the direct remote sensing of the atmospheric parameters which affect ET such as near-surface air temperature, near-surface water vapour gradients and near-surface winds. Thus, approaches for ET estimation using remotely sensed data have to work around these missing data.

Several types of remotely sensed observations may be made by measuring the electromagnetic radiation in a particular waveband reflected or emitted from the Earth's surface. The incoming solar radiation can be estimated from satellite observations of cloud cover primarily from geosynchronous orbits (Brakke and Kanemasu 1981; Tarpley 1979; Gautier *et al.* 1980). The first two groups of researchers used a regression model based on surface radiation measurements to estimate insolation, while the third used a scattering model to estimate the radiation loss in the atmosphere. For clear-sky conditions, the surface albedo may be estimated by measurements covering the entire visible and near-infrared waveband, while measurements at narrow spectral bands can be used to determine vegetative cover (Jackson 1985; Brest and Goward 1987). The surface temperature can be estimated from measurements at thermal infrared wavelengths of the emitted radiant flux, that is the 10.5 and

12.5 μm wavebands, and from some estimate of the surface emissivity (for natural surfaces, e_L is usually close to unity). Using the temperature sounders on the meteorological satellites in a linear regression model, Davis and Tarpley (1983) have estimated shelter temperatures with an error of about 2 K for clear or partly cloudy conditions. The independent variable in the linear regression is an effective surface temperature obtained by extrapolating the temperature profile from the sounder. This was compared with measured shelter temperatures to calibrate the model.

The microwave emission and reflection or backscatter from soil, primarily for wavelengths between 5 and 21 cm, are dependent on the dielectric properties of the soil which are strong functions of the soil moisture content. Thus, measurements of these microwave properties can be used to obtain estimates of the surface soil moisture. There are uncertainties in the determination of soil moisture values from microwave measurements introduced by factors such as surface roughness and vegetative cover, but it appears that microwave methods can estimate the soil moisture content in the 5 cm surface layer of the soil with four or five levels of discrimination (Schmugge et al. 1980).

The evaporation from the soil surface is directly related to the vapour pressure difference between the surface and lower atmospheric boundary layer and the water supply in the soil. Assuming that there is a suitable supply of water in the soil, the most important controlling factor is the water vapour gradient $e_a - e_s$ in equation (5.6), which is largely determined by the corresponding temperature difference.

The bare-soil evaporation can be modelled if the soil moisture can be estimated or measured. Witono and Bruckler (1988) have shown how the heat and mass flow equations (Philip and de Vries 1957) can be used to estimate evaporation from a bare soil. They used radar measurements to determine the surface soil moisture which was then used as a boundary condition for the modelling.

The presence of a plant canopy complicates the situation considerably, primarily because of the stomatal resistance terms which vary with plant type and condition. However, there is a strong relationship under certain conditions between the canopy and air temperature differences and the ET rate. This comparison of canopy and air temperatures has been used to detect crop water stress for irrigated crops in the south-western United States and these relationships have been described in a series of papers from the group at the US Water Conservation Laboratory in Phoenix, Arizona (Jackson et al. (1981) and references therein). Remotely sensed reflected solar and emitted thermal radiation has recently been used with ground-based measurements of wind, vapour pressure and incident solar radiation to estimate ET with a modified form of the Penman equation (Jackson et al. 1987). The model ET rates were compared with rates measured with Bowen ratio instrumentation over wheat, cotton and alfalfa and found to agree within about 12% and between less than 8% and 25% for daily values.

One formulation of potential ET that lends itself to remote sensing inputs is that developed by Priestley and Taylor (1972). They are obtained for saturated surfaces from

$$LE = \alpha[\Delta/(\Delta + \gamma)](R_n - G) \tag{5.7}$$

where Δ is the slope of the vapour pressure versus temperature curve, and α is an empirical evaporation constant which they determined to be 1.26. Barton (1978) and Davies and Allen (1973) have modified the result for an unsaturated surface by treating α as a function of the surface layer soil moisture. Barton used airborne microwave radiometers to sense soil moisture remotely in his study of evaporation from bare soils and grasslands. Equation (5.7) is essentially the basis for the model which Kanemasu et al. (1976) used for estimating ET with satellite data. Kotoda et al. (1983a, b) applied equation (5.7) to estimate ET using surface temperatures estimated from an airborne radiometer.

Estimates of the net radiation from geostationary satellite data were used by Heilman et al. (1977) in a Priestley–Taylor type of equation to estimate ET. The resulting moisture flux is then used to drive a water balance model for predicting crop yields. They also used Landsat multispectral data to obtain vegetation indices.

Reginato et al. (1985) combined remotely sensed reflected solar radiation and surface temperatures with ground-based meteorological data (incoming solar radiation, air temperature, wind speed and vapour pressure) to estimate the net radiation and the sensible heat flux in equation (5.1). The soil heat flux was estimated to be a fraction of the net radiation and the height of the plant canopy through the following empirical equation:

$$G = (0.1 - 0.042h)R_n \tag{5.8}$$

where h is the canopy height (m). Using the energy budget approach in equation (5.1), instantaneous values of ET were calculated for 18 wheat plots, three of which contained lysimeters. Daily ET values were estimated from the instantaneous values and compared with that calculated from soil moisture data. Figure 5.1 shows the comparison between the estimated ET values for the wettest lysimeter and Figure 5.2 compares the estimated ET and soil-moisture-derived ET for a wet, medium and dry plot (same wheat variety). Reginato et al. concluded that for clear-sky conditions, ET can be estimated adequately using a combination of ground-based and remotely sensed data, and the size of area covered appears to be limited by the distance that air temperature and wind speed data can be extrapolated. Jackson et al. (1987) extended this method by collecting the reflected solar and emitted thermal energy with low-level aircraft flights.

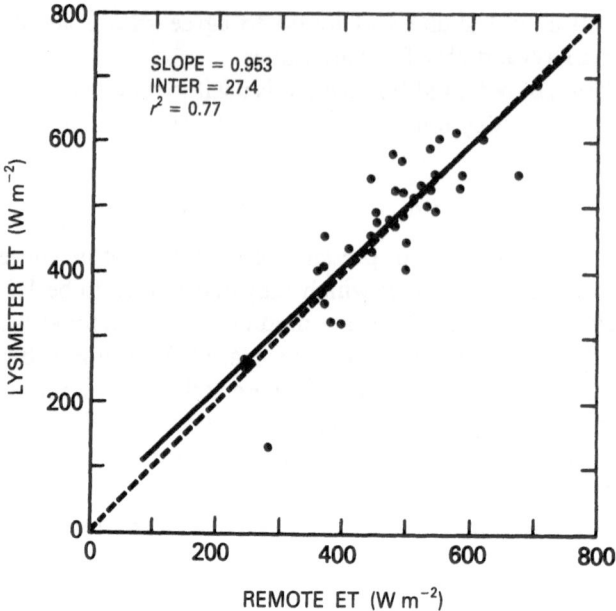

Figure 5.1 Evapotranspiration values obtained from lysimeter measurements against evapotranspiration estimated using remote inputs (after Reginato *et al.* 1985).

Traditionally, the soil heat flux would be measured by heat flux plates buried under the soil surface. However, recent research has proposed several approaches that use remote sensing data, in addition to the empirical relation shown above (equation (5.8)). Clothier *et al.* (1986) developed a relationship for estimating G from a crop height measure and a spectral vegetation index. Choudhury *et al.* (1987) proposed an empirical relationship based on a leaf area index. Kustas and Daughtry (1989) proposed a method for computing the soil heat flux based on multispectral data that should be useful for regional energy budget studies.

Incomplete canopies present an additional problem for estimating ET with remote sensing measurements. For areas with complete cover, the energy budget (equation (5.1)) can be applied by solving for the latent heat flux (Jackson 1985) and making certain assumptions about the sensible heat flux that work reasonably well (Choudhury *et al.* 1986; Huband and Monteith 1986). However, under the conditions of an incomplete canopy, there is difficulty in defining the surface temperature which is a composite of both the soil and vegetation. As a result, the resistance may differ considerably when compared with a full canopy case. This problem of using a bulk heat transfer relationship is addressed in more detail by Kustas *et al.* (1989) and Stewart *et al.* (1989). Others (e.g. Choudhury and Monteith 1988; van de Griend and

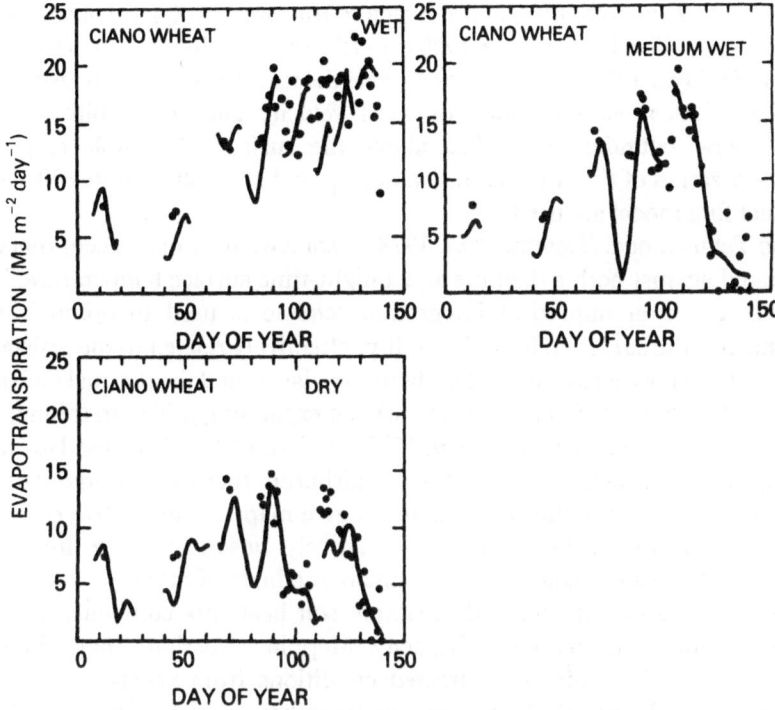

Figure 5.2 Estimates of daily evapotranspiration calculated from remote sensing (points) and from soil water content changes for Ciano wheat cultivar for three moisture conditions (after Reginato *et al.* 1985).

van Boxel 1989) have used equations (5.1)–(5.6) in multilayer models. Although these models yield good results with measured values, they may be difficult to use operationally.

A more complete but less practical approach is that adopted by Soer (1980), Rosema *et al.* (1978) and Carlson *et al.* (1981), amongst others, in which the energy balance (equations (5.1)–(5.6)) is modelled and the moisture balance is inferred from the heat flow in the soil. Equations (5.2), (5.4), (5.5) and (5.6) can be manipulated to obtain an expression for the surface energy balance in terms of the surface temperature and a set of meteorological measurements which, at least in principle, are measured or estimated routinely. The initial estimates of the soil/crop surface temperature T_s will almost certainly not satisfy the continuity requirement of equation (5.1) because of wrong initial parameter estimates, and so it is necessary to use an iterative numerical optimization technique to ensure that the condition in equation (5.1) is satisfied. The technique may be adapted to give estimates of the surface temperature T_s at every time step at which routine meteorological observations are available. The Tergra (Soer 1980) model uses a Businger–Dyer (Businger *et al.* 1971; Dyer 1967) approach to estimate the turbulent diffusion resistance, r_a. This gives r_a as a function of wind velocity and the stability of the atmospheric boundary layer just above the surface. A simple empirical parametrization of the stomatal resistance, r_s, and an explicit finite difference soil heat flux model are used.

The Tellus model (Rosema *et al.* 1978) is somewhat similar except that it is calibrated against both a daytime and a night-time surface temperature. The Newton–Raphson numerical integration scheme is used to optimize two parameters: the surface relative humidity, effectively related to the resistance to water transport across the surface boundary layer, and the thermal inertia of the soil. The finite difference scheme uses an expanding grid with thinner soil layers near the surface (Rosema *et al.* 1978). This model has been used to create lookup tables which were used with airborne thermal infrared surface temperature data and albedo estimates to give maps of cumulative daily ET (Dejace *et al.* 1979). The results are relatively close to those estimated or measured by water balance and Penman methods. Carlson *et al.* (1981) developed a similar model with a simple soil heat flux component. Using satellite data from the Heat Capacity Mapping Mission, they obtained reasonable results under non-stressed conditions from vegetated surfaces. Elkington and Hogg (1981) used an approach based on the Tergra model to estimate ET. However, they used a simpler set of inputs, estimating net radiation from the Brunt equation and the other boundary conditions from routine meteorological observations taken at nearby standard weather stations. Their results are sufficiently encouraging to suggest that the simplifications which would be required for an operational approach are indeed possible.

A procedure for estimating regional *ET* for different crops has been reported by Caselles and Delegido (1987). They use satellite-measured global radiation and maximum air temperature in the following relation:

$$ET_0 = A + BR_g + CR_g T_a \qquad (5.9)$$

where ET_0 is the maximum ET of a reference crop (green grass), R_g the global radiation obtained from satellite albedo measurements, T_a the maximum air temperature of air obtained from near-mid-day satellite temperature measurements, and A, B and C are empirical coefficients dependent upon wind and relative humidity. The calculated value of ET_0 is then modified for the specific crop in question by

$$ET_m = K_c ET_0 \qquad (5.10)$$

where ET_m is the maximum *ET* for a given crop and K_c is the crop coefficient. A schematic of this procedure is shown in Figure 5.3, and Figure 5.4 shows the results applied to the Valentian region of Spain.

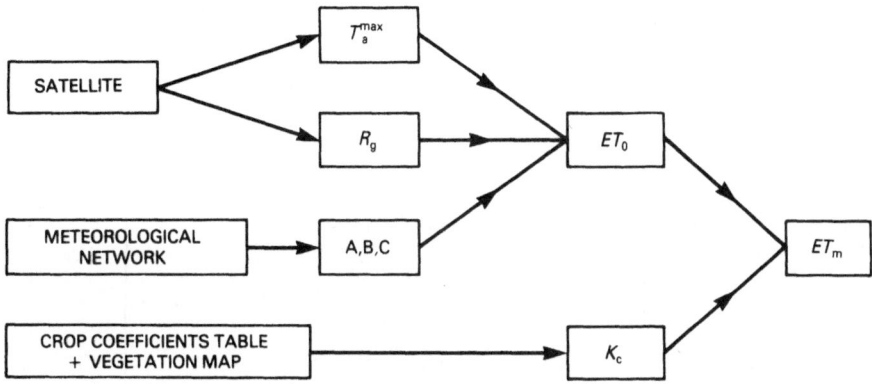

Figure 5.3 Flow diagram illustration of how remote sensing data can be integrated with meteorological and crop data to estimate evapotranspiration (after Caselles and Delegido 1987).

Bausch and Neale (1987) developed a procedure for estimating the crop coefficient from reflectance properties of a crop/soil scene. A linear transformation of the normalized difference vegetation index (Tucker *et al.* 1979) produced a seasonal curve similar to the basal crop coefficient curve. Thus, by using reflectance data we can estimate a real-time crop coefficient that has integrated the actual environmental factors such as weather, management and stress.

In another study of regional ET determination from remote sensing data, Kotoda *et al.* (1983a, b) used airborne MSS data in the 8.0–14.0 μm range to map surface temperatures for a complex land cover area in Japan. Using the Priestly–Taylor relationship (equation (5.7)), they employed the remotely sensed temperature data to estimate the Δ and R_n terms.

Price (1982) has shown how thermal data from the Heat Capacity Mapping Mission (HCMM) could be used to estimate regional-scale ET rates which were comparable with pan evaporation data. Figure 5.5 illustrates the sensitivity of the thermal data to delineated surface conditions, especially the irrigated agricultural areas that show up as cool in Figure 5.5(b). Seguin and Itier (1983) attempted to use HCMM data with a simple relationship relating daily ET to daily net radiation and a one-time surface and air temperature measurement as proposed by Jackson *et al.* (1977). They found this simple relationship did not work well for the Crau experiment (Figure 5.5). However, Seguin and Itier (1983) found that separating the data according to whether the conditions were unstable or advective would be sufficient to give good results for a medium rough surface.

Seguin *et al.* (1989) estimated evapotranspiration for the Sahel region using the relationship proposed by Jackson *et al.* (1977) and studied by Seguin and

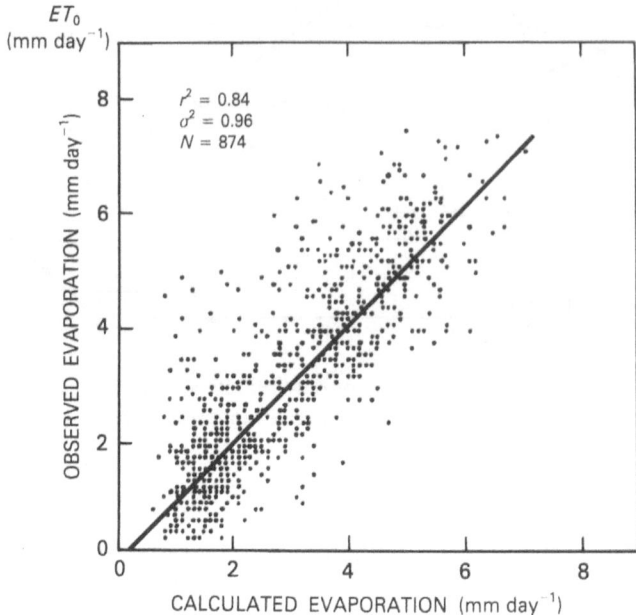

Figure 5.4 Experimental values of observed crop evapotranspiration against the calculated values obtained through an expression with the form of equation (5.8) (after Caselles and Delegido 1987).

Itier (1983) where

$$ET = R_n + A - B\,(T_s - T_a) \qquad (5.11)$$

Here ET is a daily value of evapotranspiration, R_n the net radiation, T_s the surface temperature measured by satellite and T_a the air temperature measured by a ground network. A and B are empirical constants. These results were then used to estimate and map spatially ET and millet yield for the region.

The question of how to use the spatial nature of remote sensing data to extrapolate point ET measurements to a more regional scale has been addressed by Jackson (1985) and Gash (1987). Jackson gives a good review of existing research and the problems associated with estimating the various variables in the energy balance equations. Gash has formalized an analytical framework relating the horizontal changes in evaporation to horizontal changes in surface temperature.

Bernard *et al.* (1981) examined in a simulation study the use of soil moisture estimates from a radar to calculate ET rates. The microwave estimates of the surface soil moisture were used as the upper boundary condition in a water balance model that simulated flow using the Richards equation.

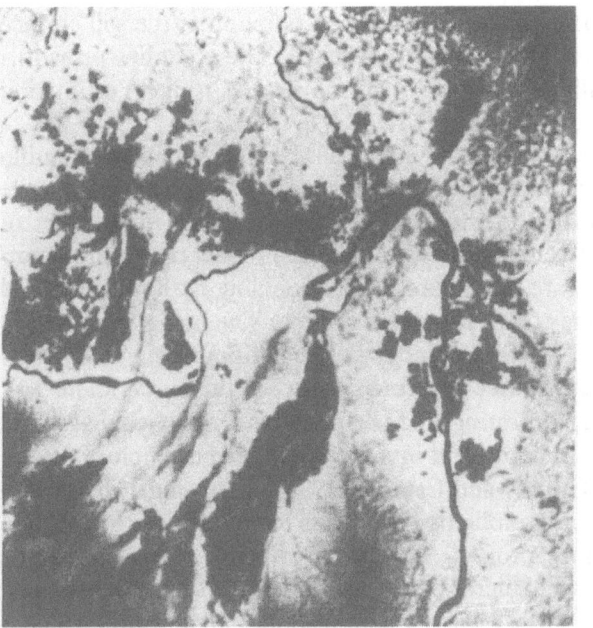

2 : 3 LT (Early Morning)

7–16°C

13 : 30 LT (Early Afternoon)

33–55°C

Figure 5.5 Early morning (a) and early afternoon (b) thermal imagery of eastern Washington State, USA. Dark areas are cold and light areas are warm. The irrigated areas are identifiable in the afternoon scene by their dark tone resulting from cooler surface temperatures (reproduced with permission from John C. Price).

Soares *et al.* (1988) used active microwave measurements of soil moisture and remotely sensed thermal data to estimate evaporation at the soil surface through an energy balance at the air/soil interface. These studies provide a good indication of the utility of microwave soil moisture sensing for estimating ET.

A more physically based and realistic model has been developed by Camillo *et al.* (1983). Both energy and moisture fluxes are modelled. The space–time dependence of temperature and moisture content is described by a set of diffusion-type partial differential equations, and the model uses a predictor/corrector method to integrate them numerically, yielding both moisture and temperature profiles in the soil as functions of time. The model was used to simulate energy and moisture fluxes under field conditions using data from Phoenix, Arizona. The qualitative agreement between the observations and the model was very good, indicating that the model is a reasonably accurate representation of the physical processes involved. A further example of the use of the model is given in Gurney and Camillo (1984). This model, because it simulates the moisture profile explicitly in addition to the temperature profile, requires an initial moisture profile. At least in principle this may also be derived using remotely sensed microwave measurements. As the model is most sensitive to changes in the surface soil temperature and soil moisture, which is that part of the profile accessible to microwave thermal infrared measurements, this is a very realistic application of remotely sensed data. If it is possible to take initial conditions from remotely sensed measurements and boundary conditions from either routine meteorological observations or remotely sensed measurements, the estimation of actual evapotranspiration from remotely sensed measurements becomes feasible.

There are various other measurements that either have been suggested as inputs to ET estimates or in principle could be used as additional model inputs. For example, multispectral information may be used to estimate the health of vegetation and the biomass (Tucker 1980; Holben *et al.* 1980) and hence to give an indication of the ET rate.

Jackson *et al.* (1985) proposed a method for calculating net radiation with remote multispectral data and ground-station meteorological data. The incoming solar and long-wave radiation are independent of surface conditions and can be measured at ground level and extended to a large area. However, the reflected solar and emitted long-wave radiation are surface dependent but measurable by remote sensing. Multispectral data were obtained from a hand-held multiband radiometer. The resulting computed net radiation values compared well with values measured with miniature net radiometers.

5.4 FUTURE APPLICATIONS

The major problems with all remote sensing methods of evapotranspiration are: (1) the process of transpiration is still not well understood and parametrized either for structured crops such as cereals or for complex vegetation such as trees; (2) in the presence of vegetation, the surface temperature T_s estimated by a thermal infrared sensor is at an unknown level within the vegetation; and (3) the most appropriate use of microwave observations of surface soil moisture in the presence of vegetation needs to be determined. For the future, we expect that the most practical method will probably use a multispectral approach including repetitive observations at the visible, near- and thermal infrared and microwave wavelengths. This will afford the possibility of estimating solar insolation, surface vegetative cover and/or albedo, surface temperature and surface soil moisture from remotely sensed data and incorporating them into models of the type described here.

Fortunately some of the space initiatives that are currently being planned should go a long way to providing the maximum amount and type of remote sensing data needed for further advances in estimating the global ET fluxes. Modellers have also been developing land phase models at a scale compatible with the needs of general circulation models (Sellers *et al.* 1986) and capable of being coupled with them. Of special interest is the Earth Observing System (Eos) being planned by NASA, ESA and Japan with major contributions being made from a number of other countries. With respect to the energy budget, the surface temperatures will be measured by MODIS, AMSU, AMSR and AMRIR (see Chapter 11). The multichannel capability of MODIS will also enable its instrumentation to make independent measurements of surface emissivity. In addition, the distribution of water vapour will be measured in varying degrees by AMSU, AMSR and LASA.

Components for determining the sensible heat flux can also be measured by the Eos instruments. As mentioned above, the temperature of the atmospheric boundary layer can be measured. Although convection and turbulence within the boundary layer cannot be measured directly, this information can be inferred from boundary layer height measurements with LASA. Also, estimates of wind speed at the top of the boundary layer can be made from cloud motions observed from geosynchronous satellites. One other component to the sensible heat flux, the surface roughness, may be able to be estimated with SAR of the land surface and with SCATT and ALT over the sea.

The latent heat flux cannot be measured directly, but Eos instrumentation will provide some sampling capability with AMSR, AMSU and LASA. Table 5.1 summarizes the potential role of the Eos instruments for estimating and measuring components useful for the determination of the energy and water vapour fluxes.

Looking at the land phase of ET, it can be seen that the capabilities for measuring solar radiation from GOES, albedo from MODIS and HIRIS, and surface temperatures from TIMS, MODIS and GOES will provide valuable information for such models as the Priestley–Taylor (equation (5.7)) relationships. Plant conditions themselves may be able to be measured and monitored through thermal data with TIMS, MODIS and GOES, changes in

Table 5.1 Eos instruments for determining energy fluxes at the Earth's surface (see also Table 11.1)

Radiative energy input

Direct measurements
    Solar flux at top of atmosphere   SCM, SUSIM
    Cloud cover   MODIS, VAS★
    Aerosols   MODIS, LASA
    Surface albedo   MODIS, HIRIS, AMRIR
Indirect measurements to determine energy
transmitted through clouds
    Cloud temperature and height   MODIS, LASA, VAS★
    Cloud total liquid and ice content   AMSR
    Cloud vertical extent and number of layers   AMSU, AMRIR, MODIS, VAS★

Radiative energy lost from surface

Surface temperature   MODIS, AMSU, AMSR,
         AMRIR, TIMS, VAS★
Surface emissivity   MODIS, AMSU, AMSR
Fractional sea ice   AMSR, SAR
Greenhouse gases
    CO_2 Ground-based monitoring
    Water vapour   AMSU, AMSR, LASA, MODIS,
         VAS★
    Ozone   GOMR, IR-Rad, MLS, SUB-MM,
         CIS
    CH_4 (troposphere)   NCIS, CR (maybe)
    CH_4, N_2O, chlorofluorocarbons   IR-Rad, SUB-MM, CIS, GOMR
        (stratosphere)

Sensible heat flux

Direct measurements   (see above)
Indirect measurements
    Temperature of atmospheric boundary layer   LASA
        based on boundary layer height
    Tropospheric temperature profile   MODIS, AMSU, VAS★, LASA
    Intensity of convection in boundary layer
        based on:
        Wind at top of boundary layer   LAWS, VAS★, cloud track winds★,
             atmospheric temperature
        Surface winds (ocean)   SCATT, ALT
        Surface roughness: land   SAR
                 ocean   SCATT, ALT

Latent heat flux

Direct measurements
    Atmospheric water vapour   (see above)
    Atmospheric liquid water and ice   AMSR
Indirect measurements
    Ocean surface evaporation   AMSR, AMSU, MODIS, AMRIR
    Land surface evaporation   HIRIS, MODIS, SAR, ESTAR,
         TIMS

★ Geostationary platform instruments.

reflectance caused by stress through MODIS and HIRIS, and changes in geometry and dielectric constant (moisture content) with SAR. Various spaceborne radars will also be operating in the same time period and will give additional useful information on soils and vegetation.

In summary, future programmes such as Eos and the multispectral capabilities of the instrumentation should provide unusual and exciting new data for evaluating evaporation and transpiration on local, regional and global scales, the study of which has so far been limited much more to the laboratory or to special experiments.

REFERENCES

Barton, I. J. (1978) A case study comparison of microwave radiometer measurements over bare and vegetated surfaces. *J. Geophy. Res.* **83**, 3513–17.

Bausch, W. C. and Neale, C. M. U. (1987) Crop coefficients derived from reflected canopy radiation: a concept. *Trans. Am. Soc. Agric. Eng.* **30**, 701–9.

Bernard, R., Vauclin, M. and Vidal-Madjar, D. (1981) Possible use of active microwave remote sensing data for prediction of regional evaporation by numerical simulation of soil water movement in the unsaturated zone. *Water Resour. Res.* **17**, 1603–10.

Brakke, T. W. and Kanemasu, E. T. (1981) Insolation estimation from satellite measurements of reflected radiation. *Remote Sensing Environ.* **11**, 157–67.

Brest, C. L. and Goward, S. N. (1987) Deriving surface albedo measurements from narrow band satellite data. *Int. J. Remote Sensing* **8**, 351–67.

Brown, K. W. and Rosenberg, N. J. (1973) A resistance model to predict evapotranspiration and its application to a sugar beet field. *Agron. J.* **65**, 341–2.

Brutsaert, W. H. (1982) *Evaporation into the Atmosphere: theory, history and application*, Reidel, Boston, MA.

Businger, J. A., Wyngaard, J. C., Izumi, Y. and Bradley, E. F. (1971) Flux profile relationships in the atmospheric surface layer. *J. Atmos. Sci.* **28**, 181–9.

Camillo, P., Gurney, R. J. and Schmugge, T. J. (1983) A soil and atmospheric boundary layer model for evapotranspiration and soil moisture studies. *Water Resour. Res.* **19**, 371–80.

Carlson, T. N., Dodd, J. K., Benjamin, S. G. and Cooper, J. N. (1981) Remote estimation of surface energy balance, moisture availability and thermal inertia over urban and rural areas. *J. Appl. Meteorol.* **20**, 67–87.

Caselles, V. and Delegido, J. (1987) A simple model to estimate the daily value of the maximum evapotranspiration from satellite temperature and albedo images. *Int. J. Remote Sensing* **8**, 1151–62.

Choudhury, B. J., Idso, S. B. and Reginato, R. J. (1987) Analysis of an empirical model for soil heat flux under a growing wheat crop for estimating evaporation by an infrared-temperature based energy balance equation. *Agric. Forest Meteorol.* **39**, 283–97.

Choudhury, B. J. and Monteith, J. L. (1988) A four-layer model for the heat budget of homogeneous land surfaces. *Q. J. R. Meteorol. Soc.* **114**, 373–98.

Choudhury, B. J., Reginato, R. J. and Idso, S. B. (1986) An analysis of infrared temperature observations over wheat and calculation of the latent heat flux. *Agric. Forest Meteorol.* **37**, 75–88.

Clothier, B. E., Clawson, K. L., Pinter, Jr, P. J., Moran, M. S., Reginato, R. J. and Jackson, R. D. (1986) *Agric. Forest Meteorol.* **37**, 319–29.

Davies, J. A. and Allen, C. D. (1973) Equilibrium, potential and actual evaporation from cropped surfaces in southern Ontario. *J. Appl. Meteorol.* **12**, 649–57.

Davis, P. A. and Tarpley, J. D. (1983) Estimation of shelter temperatures from operational satellite sounder data. *J. Clim. Appl. Meteorol.* **22**, 369–76.

Dejace, J., Megier, J., Kohl, M., Maracci, G., Reiniger, P., Tassone, G. and Huygen, J. (1979) Mapping thermal inertia, soil moisture and evaporation from aircraft day and night thermal data. *13th ERIM Symp. on Remote Sensing, Manila*, pp. 1015–24.

Dyer, A. J. (1967) The turbulent transport of heat and water vapor in an unstable atmosphere. *Q. J. R. Meteorol. Soc.* **93**, 501–8.

Elkington, M. D. and Hogg, J. (1981) The characteristics of soil moisture content and actual evapotranspiration from crop canopies using thermal infrared remote sensing. *Proc. Remote Sensing Soc., Reading.*

Gash, J. H. C. (1987) An analytical framework for extrapolating evaporation measurements by remote sensing surface temperature. *Int. J. Remote Sensing* **8**, 1245–9.

Gautier, C., Diak, G. and Masse, S. (1980) A simple model to estimate incident solar radiation at the surface from GOES satellite data. *J. Appl. Meteorol.* **19**, 1005–11.

Gurney, R. J. and Camillo, P. J. (1984) Modelling daily evaporation using remotely-sensed data. *J. Hydrol.* **69**, 305–24.

Heilman, J. L., Kanemasu, E. T., Bagley, J. O. and Rasmussen, V. P. (1977) Evaluating soil moisture and yield of winter wheat in the Great Plains using Landsat data. *Remote Sensing Environ.* **6**, 315–26.

Holben, B. N., Tucker, C. J. and Fan, C. J. (1980) Spectral assessment of soybean leaf area and leaf biomass. *Photogramm. Eng. Remote Sensing* **46**, 651–6.

Huband, N. D. S. and Monteith, J. L. (1986) Radiative surface temperature and energy balance of a wheat canopy. *Boundary Layer Meteorol.* **36**, 1–17.

Jackson, R. D. (1985) Evaluating evapotranspiration at local and regional scales. *IEEE Trans. Geosci. Remote Sensing* **GE-73**, 1086–95.

Jackson, R. D., Idso, S. B., Reginato, R. J. and Pinter, P. J. (1981) Canopy temperature as a crop water stress indicator. *Water Resour. Res.* **17**, 1133–8.

Jackson, R. D., Moran, M. S., Gay, L. W. and Raymond, L. H. (1987) Evaluating evaporation from field crops using airborne radiometry and ground-based meteorological data. *Irrig. Sci.* **8**, 81–90.

Jackson, R. D., Pinter, Jr, P. J. and Reginato, R. J. (1985) Net radiation calculated from remote multispectral and ground station meteorological data. *Agric. Forest Meteorol.* **35**, 153–64.

Jackson, R. D., Reginato, R. J. and Idso, S. B. (1977) Wheat canopy temperature: a practical tool for evaluating water requirements. *Water Resour. Res.* **13**, 651–6.

Jensen, M. E. and Haise, H. R. (1963) Estimating evapotranspiration from solar radiation. *Proc. American Society of Civil Engineering, J. Irrig. Drain. Div.* **89**, 15–41.

Kanemasu, E. T., Stone, L. R. and Powers, W. L. (1977) Evapotranspiration model tested for soybeans and sorghum. *Agron. J.* **68**, 569–72.

Kotoda, K., Nakagawa, S., Kai, K., Yoshino, M. M., Takeda, K. and Seki, K. (1983a) Application of equilibrium evaporation model to estimate evapotranspiration by remote sensing technique in remote sensing of snow and evapotranspiration, Part II. *Proc. 2nd US/Japan Workshop on Remote Sensing of Snow and Evapotranspiration, Honolulu, Hawaii*, NASA CP-2363, pp. 115–23.

Kotoda, K., Nakagama, S., Kai, K., Yoshino, M. M., Takeda, K. and Seki, K. (1983b) Estimation of regional evapotranspiration using remotely sensed land surface temperature. *Proc. 2nd US/Japan Workshop on Remote Sensing of Snow and Evapotranspiration, Honolulu, Hawaii*, NASA CP-2363, pp. 99–114.

Kustas, W. P., Choudhury, B. J., Moran, M. S., Reginato, R. J., Jackson, R. D., Gay, L. W. and Weaver, H. L. (1989) Determinations of sensible heat flux over sparse canopy using thermal infrared data. *Agric. Forest Meteorol.* **44**, 197–216.

Kustas, W. P. and Daughtry, C. S. T. (1989) Estimation of the soil heat flux/net radiation ratio from spectral data. *Agric. Forest Meteorol.* **49**, 205–33.

Monteith, J. L. (1973) *Principles of Environmental Physics*, Edward Arnold, London.

Monteith, J.L. (1981) Evaporation and surface temperature. *Q. J. R. Meteorol. Soc.* **107**, 1–27.

Penman, H. L. (1948) Natural evaporation from open water, bare soil and grass. *Proc. R. Soc.* A **193**, 129–45.

Philip, J. R. and De Vries, D. A. (1957) Moisture movement in porous materials under temperature gradients. *Eos Trans. Am. Geophys. Union* **30**, 222–32.

Price, J. C. (1982) Estimation of regional scale evapotranspiration through analysis of satellite thermal-infrared data. *IEEE Trans. Geosci. Remote Sensing* **GE-20**, 286–92.

Price, J. C. (1982) On the use of satellite data to infer surface fluxes at meteorological scales. *J. Appl. Meteorol.* **21**, 1111–22.

Priestley, C. H. B. and Taylor, R. J. (1972) On the assessment of surface heat flux and evaporation using large-scale parameters. *Mon. Weather Rev.* **100**, 82–92.

Reginato, R. J., Jackson, R. D. and Pinter, Jr, P. J. (1985) Evapotranspiration calculated from remote multispectral and ground station meteorological data. *Remote Sensing Environ.* **18**, 75–89.

Rosema, A., Bijleveld, J. H., Reiniger, P., Tassone, G., Gurney, R. J. and Blyth, K. (1978) Tellus, a combined surface temperature, soil moisture and evaporation mapping approach. *Proc. 12th ERIM Symp. on Remote Sensing, Ann Arbor, MI*, p. 10.

Schmugge, T. J., Jackson, T. J. and McKim, H. L. (1980) Survey of methods for soil moisture determination. *Water Resour. Res.* **16**, 961–79.

Seguin, B., Assad, E., Freteaud, J. P., Imbernon, J., Kerr, Y. and Lagouarde, J.P. (1989) Use of meteorological satellites for water balance monitoring in Sahelian regions. *Int. J. Remote Sensing* **10**, 1101–17.

Seguin, B. and Itier, B. (1983) Using midday surface temperature to estimate daily evapotranspiration from satellite thermal IR data. *Int. J. Remote Sensing* **4**, 371–83.

Sellers, P. J., Mintz, Y., Sud, Y. C. and Dalcher, A. (1986) A simple biosphere model (SiB) for use with general circulation models. *J. Atmosp. Sci.* **43**, 505–31.

Soares, J. V., Bernard, R., Taconet, O., Vidal-Madjar, D. and Weill, A. (1988) Estimation of bare soil evaporation from microwave measurements. *J. Hydrology* **99**, 281–96.

Soer, G. J. R. (1980) Estimation of regional evapotranspiration and soil moisture conditions using remotely sensed crop surface temperatures. *Remote Sensing Environ.* **9**, 27–45.

Stewart, J. B., Shuttleworth, W. J., Blyth, K. and Lloyd, C.R. (1989) *Preprints, 19th Conf. Agricultural and Forest Meteorology and 9th Conf. Biometeorology and Aeriobiology, Charleston, SC*, pp. 144–6.

Tarpley, J. D. (1979) Estimating incident solar radiation at the surface from geostationary satellite data. *J. Appl. Meteorol.* **18**, 1172–81.

Thornthwaite, C. W. (1948) An approach toward a rational classification of climates. *Geophys. Rev.* **38**, 55–94.

Tucker, C. J. (1980) Remote sensing of leaf water content in the near-infrared. *Remote Sensing Environ.* **10**, 23–32.

Tucker, C. J. (1983) Remote sensing of soil moisture with microwave radiometers. *Trans. Am. Soc. Agric. Eng.* **26**, 748–53.

Tucker, C. J., Elgies, J. H., McMurtrey, III, J. E. and Fan, C. J. (1979) Monitoring corn and soybean crop development with hand-held radiometric spectral data. *Remote Sensing Environ.* **8**, 237–48.

van de Griend, A. A. and van Boxel, J. H. (1989) Water and surface energy balance model with a multilayer canopy representation for remote sensing purposes. *Water Resour. Res.* **25**, 949–71.

Witono, H. and Bruckler, L. (1988) Use of microwave backscatter measurements as boundary conditions, in *Soil/Atmosphere Water Transfer Modelling*. Proc. 4th International Colloq. Spectral Signatures of Objects in Remote Sensing, Aussois, France, ESA-SP-289, pp. 37–42.

6

Runoff

6.1 INTRODUCTION

Runoff is the one hydrologic variable that is most often used by hydrologists and water resource planners. Accurate prediction of runoff rates and runoff volumes is used for water supply forecasting, flood predictions and warnings, navigation, water quality management, hydropower production, and many other water resource applications. The objective most sought by hydrologists is the accurate and timely prediction of runoff at a given point in a drainage basin. The tools available to hydrologists encompass a wide range of equations and models as well as stream gauging stations where the streamflow or reservoir volume is measured directly in real time.

The limitation that hydrologists work under with respect to the analytical tools available, that is the equations and models, is that none of these work well under all conditions. The general objective in hydrologic research has been to develop better models that have more general applications. Direct measurement of runoff provides excellent and timely data but it is limited in use to the exact location where it was collected. Stream measuring stations are more common in large river systems and reservoirs where day-to-day management is a goal and the stations can be economically justified.

Hydrologic data and information used as input to the various models is another need that hydrologists must satisfy. Many models have been designed to use only readily available data such as published maps and soil surveys. Deterministic or special-purpose models need more detailed data than are usually available. Thus, costly and time-consuming field data must be collected for the basic inputs, a step which often rules out the use of some of the more complex but physically based models. This chapter will illustrate how remotely sensed data are currently being used by hydrologists. The future potential for remote sensing applications to hydrology will also be discussed.

6.2 GENERAL APPROACH

Runoff cannot be directly measured by remote sensing techniques. The role of remote sensing in runoff calculations is generally to provide a source of input data or as an aid for estimating equation coefficients and model parameters. Satellite communications have been an extremely useful means for transmitting measured streamflow data to a central point; however, this is not strictly a remote sensing application (Shope and Paulson 1986).

There are two general areas where remote sensing has currently been used as input data for computing runoff. The first is based on producing input data for a class of empirical flood peak, annual runoff, or low-flow equations. These relationships are based on various geomorphic descriptions of a basin and on whether remote sensing is used in the same way as aerial photography. In the second approach, runoff models that are based on a land use component have been modified to use digital analysis or image interpretation of multispectral data to delineate land use classes.

6.3 CURRENT APPLICATIONS

Empirical relationships

Empirical flood formulae can be useful for making quick estimates of peak flow or for making these estimates when there is very little other information available. A user must be fully aware, however, of their limitations. Generally these equations are restricted to a size range of the basin and the climatic/hydrologic region of the world in which they were developed.

Most of the empirical flood formulae are based on the drainage area of the basin and take one of several forms; that is, from the very simple

$$Q = cA^n \tag{6.1}$$

to the more complex

$$Q = \frac{cA}{a + bA}m + dA \tag{6.2}$$

where Q is the peak flow rate, A is the area of the basin, and a, b, c, d, m and n are coefficients or exponents which must be evaluated for a specific region and range of basin areas. A list of empirical flood formulae from many regions of the world can be found in the United Nations Flood Control Series No. 7 (United Nations 1955).

Landsat data can be used to improve empirical equations of various runoff characteristics. Regression equations relating runoff to basin characteristics have been proposed by a number of hydrologists, including Thomas and Benson (1970). More recent work by Allord and Scarpace (1979) has shown how the addition of Landsat data can improve regression equations based on topographic maps alone. The addition of land cover determined from Landsat reduced the standard estimate of error by 9% for 2 and 10 year 7-day low flow values, and by 14% for the 10, 50 and 100-year flood frequency estimates. In the regression analysis, four of the nine basin characteristics determined from Landsat data were significant, whereas only one of 12 basin characteristics developed from topographic maps was significant.

In another application, mean annual and mean seasonal (monsoon) discharge equations were developed by Chandra and Sharma (1978) from basin characteristics measured from Landsat scenes. Their equations, which are empirical and valid only for the Ramganga basin, are as follows:

$$Q_m = -562\ 226 + 328\ 157D_d + 2612.68S - 51.69EL + 32\ 616R_c$$
$$-117.98A - 6158.6FP - 4593.9FN + 1.553P \qquad (6.3)$$

and

$$Q_a = 93\ 583.3 + 0.901P - 3958.5FP + 8417SP - 692.5AGP \quad (6.4)$$

where Q_m and Q_a are the monsoon and annual discharges, respectively, D_d is the drainage density, S is the slope, EL is the elevation in meters, R_c is the circulatory ratio, A is the basin area in square kilometres, FP is the forest area in square kilometres, FN is the equivalent non-forest area, P is the rainfall in inches, SP and AGP are the land areas in shrubs and agriculture, respectively. This type of empirical equation can be very useful if used only in the basins where they were derived. Similar equations can be developed if the data are readily available; however, users should be acutely aware of the limitations, which are great.

Other geomorphic characteristics have been used by hydrologists as estimates of various flood statistics. For example, Inglis (1939) proposed an equation relating river meander length to an approximation of the 100-year flood, as follows:

$$Q_{100} = (ML)C^2 \qquad (6.5)$$

where C is a coefficient with values ranging between 18 and 42, and ML is the meander length in feet which can be scaled from aerial photography, Landsat imagery, or other remotely sensed data. Undoubtedly, there are a number of other runoff relationships that use map measurements for their variables which can be adapted to use the remotely sensed data in a similar way.

Watershed geometry

Remote sensing data can be used to obtain almost any information that is typically obtained from maps. In many regions of the world, remotely sensed data, and particularly Landsat or SPOT data, may be the only source of good cartographic information. Drainage basin area and the drainage network are easily obtained from good imagery, even in remote regions. There have also been a number of studies to extract quantitative geomorphic information from Landsat imagery (Haralick *et al.* 1985). Once basic measurements have been taken from the imagery, the usual physiographic descriptors can be calculated, such as basin shape, circularity and stream orders.

Selection of imagery is important if the maximum possible information is to be obtained. Vegetation state is an important consideration. In the case of Landsat or SPOT, the choice of imagery with a low sun angle will enhance topographic and drainage features. Landsat MSS bands 5 (0.6–0.7 μm) and 7 (0.8–1.1 μm) and TM bands 3 (0.63–0.69 μm), 4 (0.76–0.9 μm) and 5 (1.55–1.75 μm) have proven to be the best choices for discerning physiographic features. The visible red band (MSS band 5, TM band 3) is best for showing stream channel networks when their size is too small to be detected directly. This band is also good for separating vegetation types and for delineating non-vegetated areas. MSS band 7 shows the most contrast between water and land areas. In tropical regions, side-looking airborne radar (SLAR) can penetrate the dense vegetation and produce an image that exhibits topographic features and drainage patterns.

Recent use of the SPOT imagery and its stereographic capabilities have demonstrated its potential in topographic mapping. Gugan and Dowman (1988) have demonstrated a geometric model for image restitution that can produce topographic maps at 1 : 50 000 scale and 25 m contours.

The enhanced spatial resolution available from TM and SPOT data allows significantly greater geomorphic data to be obtained than from the Landsat MSS. France and Hedges (1986) compared MSS and TM data for their application to hydrology and found that the TM data provided much more information. Table 6.1 shows a comparison of channel lengths and drainage

Table 6.1 Comparison of channel lengths and drainage densities discernible from MSS, TM, photographs and a map for an area of 235 km^2 in North Wales, UK (from France and Hedges, 1986)

	MSS	TM	Photos	Map
Total channel length (km)	25.21	88.75	121.50	272.24
Drainage density (km/km^2)	0.11	0.38	0.52	1.15

densities discernable from TM and MSS data when compared to black and white photos and a 1 : 50 000 scale map.

Ungar *et al.* (1988) have described a technique for rotating a nadir SPOT pass so that it can be aligned with an off-nadir pass. Elevations for each nadir pixel were determined from the displacements of the relative positions of the nadir and off-nadir pixels and the off-nadir pass look angle. This technique can reliably provide a 10 m vertical resolution field at 30 m spacing with the 10 m SPOT HRV panchromatic data.

6.4 RUNOFF MODELS

To date, most uses of satellite remote sensing data have consisted of relatively straightforward extensions of photogrammetry. There have been a number of successful applications where Landsat data have been used to determine both urban and rural land use for estimating runoff coefficients.

Land use and runoff coefficients

Land use is an important characteristic of the runoff process that affects infiltration, erosion and evapotranspiration. Thus, almost any physically based hydrologic model uses some form of land use data or parameters based on these data. Distributed models, in particular, need specific data on land use and their location within the basin. Some of the first research for adapting satellite-derived land use data was done by Jackson *et al.* (1976) with the US Army Corps of Engineers STORM model (US Army Corps of Engineers 1976). However, most of the work on adapting remote sensing to hydrologic modelling has been with the Soil Conservation Service (SCS) runoff curve number (RCN) model (US Department of Agriculture 1972). The SCS models are among the most widely used in hydrology and water resource planning of agricultural areas in the United States. These models were originally developed for predicting runoff volumes from agricultural fields and small watersheds. However, they have been expanded for subsequent use in a wide variety of conditions at many basin sizes including urban and suburban areas.

The storm treatment overflow runoff model (STORM) (US Army Corps of Engineers 1976) is an urban hydrology model that has been used for hydrologic and water quality studies and planning. STORM is a continuous simulation model that requires hourly rainfall data and a relatively simple runoff coefficient to partition rainfall into runoff. The STORM model is based on

$$R = CA(P - D) \tag{6.6}$$

where R is the runoff volume, A the drainage basin area, P the rainfall volume, D the depression storage, and C the runoff coefficient expressed as the fraction of rainfall that becomes runoff.

The runoff coefficient can be expressed as an empirical relationship based on the percentage of imperviousness (IMP) and separate runoff coefficients for the impervious areas (C_i) and the previous areas (C_p), according to

$$C = C_p (1 - IMP) + C_i(IMP) \qquad (6.7)$$

Usually these runoff coefficients are determined by calibration with rainfall and runoff data and the percentage of imperviousness is measured by land use surveys or aerial photograph delineation. Often the percentage imperviousness is also treated as a parameter and altered, together with the runoff coefficients, to achieve a good match between the simulated and measured runoff. Jackson *et al.* (1976) modified the procedures to be more amenable for use with digitally classified Landsat data. They used data from several watersheds to determine C_i and C_p empirically and they defined the maximum depression storage, D_{max}, as

$$D_{max} = 0.25 (1 - IMP) + 0.06 (IMP) \qquad (6.8)$$

where the coefficients 0.25 and 0.06 are the maximum available depths of depression storage for pervious and impervious areas, respectively, as suggested by Tholin and Keifer (1959). In the application by Jackson *et al.* (1977), IMP was determined directly from analysis of Landsat data. They evaluated the success of this approach by comparing the results of the model using Landsat data with those of the model used in the conventional way. Table 6.2 shows a comparison between the measured storm discharges and those discharges simulated using the conventional approach and the Landsat approach.

It is obvious from Table 6.2 that the Landsat approach can yield results as good as the conventional approach. This point is reinforced in Figure 6.1

Table 6.2 Comparison of observed peak discharges with STORM simulated peak discharges using conventional and Landsat methods (Jackson *et al.* 1977)

Event data	Discharge ($ft^3 s^{-1}$)		
	Observed	*Conventional*	*Landsat*
20 August 1963	11 700	11 952	11 192
14 September 1966	6900	7981	7473
22 July 1969	14 600	14 506	13 939
2 August 1970	8300	9378	8782
9 July 1970	8800	7881	7379
22 June 1972	10 000	7273	6815

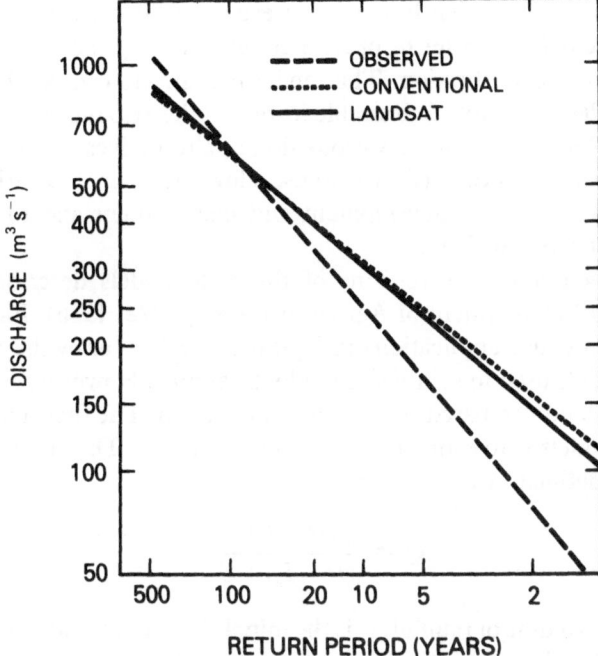

Figure 6.1 Illustration of Landsat-derived flood frequency curves compared with Fourmile Run gauged watershed (1 cubic foot per second (cfs) = 0.0283 m³ s⁻¹)(after Jackson *et al.* 1977).

where the Landsat-simulated flood frequency curve is practically identical to the conventionally simulated curve.

The results of this study (Jackson *et al.* 1977) indicate that for planning studies the Landsat approach is highly cost effective. The authors estimated that the cost benefits were of the order of 2.5 to 1 or 6 to 1, in favour of the Landsat approach, depending on the experience of the analysts and the availability of data and background information. These benefits would increase even more for larger basins or for multiple basins in the same general hydrologic area.

In another urban hydrology application, Draper and Rao (1986) used a single linear reservoir hydrograph model with a precipitation excess equation based on the percentage of impervious area, storm rainfall, area covered by ground moraine deposits, and an antecedent rainfall index. Two computerized methods for determining the impervious area were demonstrated. Using high-altitude aerial photography, a laser imaging processing scanner and a videotape camera system were compared favourably with conventional methods of aerial photograph interpretation.

The empirical SCS models have widespread appeal because the major input parameters are defined in terms of land use and soil type and do not require hydrologic data for calibration. The model parameters can be chosen from handbook tables and maps. A desirable feature of the SCS models is the ability for the hydrologist to simulate various design alternatives and compare the results. The parameter defined by land use allows the user to experiment with alternative forms of land development and management and to assess the impact of the proposed changes.

There are a number of versions of the SCS models described in SCS documents (US Department of Agriculture 1969, 1972, 1986). Each version differs by its intended application (agricultural or urban) or by the complexity of the submodels used to synthesize the hydrographs. However, each version uses the same land-use-based runoff volume equation. The first step in the use of the SCS models is to estimate the volume of runoff. The volume of direct runoff, Q, is defined by the equation

$$Q = \frac{(P - I_a)^2}{(P - I_a) + S} \qquad (6.9)$$

where P is the volume of rainfall, I_a is the initial abstraction and S is a retention parameter defined as the sum of I_a and the potential maximum retention of the watershed. Conceptually, I_a represents the interception, infiltration and depression storage that must be satisfied before runoff begins. In the SCS model, the initial abstraction is defined as

$$I_a = 0.2\,S \qquad (6.10)$$

which is based on an analysis of small watershed rainfall–runoff data. Substituting equation (6.10) into equation (6.9) yields the basic SCS model runoff equation

$$Q = \frac{(P - 0.2S)^2}{(P + 0.8S)} \qquad (6.11)$$

In practice, a runoff curve number (RCN) is defined as a transformation of S according to the relationship

$$RCN = 1000/(S + 10) \qquad (6.12)$$

The RCN is a coefficient developed from one of four hydrologic soil groups which has been adjusted for land use and management practice. The RCN may be further modified by an antecedent precipitation index to account for very wet or dry conditions.

Several studies have been conducted to demonstrate the feasibility of developing land use categories from remotely sensed data. At first urban and suburban areas were studied because the greatest contrast would be available

Table 6.3 Comparison of aerial photograph and Landsat-derived discharges computed with the SCA runoff curve number model (from Ragan and Jackson 1980)

Return period (years)	Precipitation (inches)	Discharge $(ft^3 s^{-1})$	
		Aerial photographs	Landsat
2	3.0	3850	3490
5	3.3	6064	5140
10	5.4	7580	6900
25	5.8	9300	8759
50	6.7	10 400	9900
100	7.3	11 806	11 100

between the impervious and other more pervious areas. In early work with remotely sensed data, Jackson et al. (1977) demonstrated that land cover (particularly the percentage of imperviousness) could be used effectively in the US Army Corps of Engineers STORM model (US Army Corps of Engineers, 1976). In an extension of the same study, Jackson and Ragan (1977) used Bayesian decision theory to demonstrate that computer-aided analysis of Landsat data was highly cost effective. In a study of the upper Anacostia River basin in Maryland, Ragan and Jackson (1980) demonstrated that Landsat-derived land use data could be used for calculating synthetic flood frequency relationships. The Landsat-derived results were compared with predicted values developed by conventional procedures that used low-level aerial photography to determine the land use. Table 6.3 is a comparison of these two procedures.

The important thing to note is that the two results are very nearly equal, especially when the flow rates or the flow depths are compared. This is a most important aspect of remote sensing applications. The land use statistics are seldom duplicated but the hydrologic results are no less acceptable. Another study of three complex basins in Pennsylvania demonstrates this point (Bondelid et al. 1982). Table 6.4 compares the land use statistics and the computed runoff curve numbers. It can easily be seen that the individual watersheds' land use statistics may differ by a large amount but the computed curve numbers are nearly the same. A study by the US Army Corps of Engineers (Rango et al. 1983) estimated that any individual pixel may be incorrectly classified about one-third of the time. However, by aggregating land use over a significant area, the misclassification of land use can be reduced to about 2%. These results are all based on studies that used Landsat

Table 6.4 Summary of conventional versus Landsat land cover classifications and comparison of curve numbers for the Little Mahoney and Chickies Creek watersheds (Bondelid *et al.* 1982)

Land cover (%)	Conventional	Landsat
Little Mahoney Creek		
Forest	76.0	82.3
Agriculture	6.9	0.1
Urban	16.0	14.3
Curve number	72.0	70.0
Chickies Creek		
Forest	29.0	30.5
Agriculture	57.0	62.4
Urban	14.0	7.1
Curve number	73.0	73.0

multispectral scanner data (80 m resolution). Even better results can be expected with 30 m thematic mapper data from Landsat-4 or 20 m data from the SPOT satellite.

The resolution of the Landsat MSS data is about 1 acre (80 m) and this does not provide the detail necessary to use the published SCS land use tables (see Chapter 9, US Department of Agriculture, 1972). Consequently, we must develop a land cover table analogous to Table 2.2a, 2.2b, 2.2c or 2.2d of US Department of Agriculture (1986) but compatible with Landsat data. Such tables have much less detail than the published SCS land use tables. Tables 6.5, 6.6 and 6.7 here are examples of land use categories developed for Landsat data. Each of these is specific to the particular area where the study was conducted. The RCN relationship developed by different workers for using Landsat data show some inconsistencies in the RCN values selected for similar land uses. This is probably the result of individual differences in how land use was defined, the training areas chosen, and the characteristics of the study area. To date, no general land use table or classification for remotely sensed data has been developed. We must develop a table suitable for the specific watershed and design a classifier to delineate the chosen land use areas.

Still and Shih (1985) used Landsat data to develop a basin-wide runoff index which was compared with an earlier index determined from USGS land use maps. They concluded that the runoff index had not changed but in so doing have demonstrated how remotely sensed data can be used to track the changes in runoff that occur in a basin due to land use changes.

Table 6.5 Runoff curve numbers for Landsat land use delineations (from Ragan and Jackson 1980)

Land use description	Hydrologic soil group			
	A	*B*	*C*	*D*
Forest land	25	55	70	77
Grassed open space	36	60	73	78
Highly impervious (commercial–industrial–parking lot)	90	93	94	94
Residential	60	74	83	87
Bare ground	72	82	88	90

Table 6.6 Runoff curve numbers for Landsat rural land cover delineations (from Slack and Welch 1980)

Land use description	Hydrologic soil group			
	A	*B*	*C*	*D*
Agricultural, vegetated	52	68	79	84
Bare ground	77	86	91	94
Forest land	30	58	72	78

Table 6.7 Pennsylvania Landsat runoff curve numbers (from Bondelid *et al.* 1982)

Land cover	Hydrologic soil group			
	A	*B*	*C*	*D*
Woods	25	55	70	77
Agriculture	64	75	83	87
Residential	60	74	83	87
Highly impervious	90	93	94	95
Water	98	98	98	98

Mauser (1984) used the SCS TR-20 model for calculating flood hydrographs for the Dreisam watershed in the southern Black Forest in West Germany. A geographic database incorporated satellite-determined land use, soil information and slope information at a resolution of 64 × 104 m. The land use was determined from MSS data with a maximum-likelihood classification scheme. With this approach, Mauser was able to reproduce the flood frequency history of the basin. He also demonstrated how this approach can be expanded to simulate possible hydrologic changes brought about by land use changes. Figure 6.2 illustrates the results of potential land use changes brought about by hypothetical deforestation caused by air pollution.

There have been several attempts to measure the SCS curve number directly with direct measures of reflectance. With this approach it is implicitly assumed that the reflectance is somehow related to the general land use which is an integrated land use of the basin. Thus, rather than determining an area of each land use type or assigning a curve number to each pixel, a mean reflectance value for the entire basin or combination of different band reflectances is

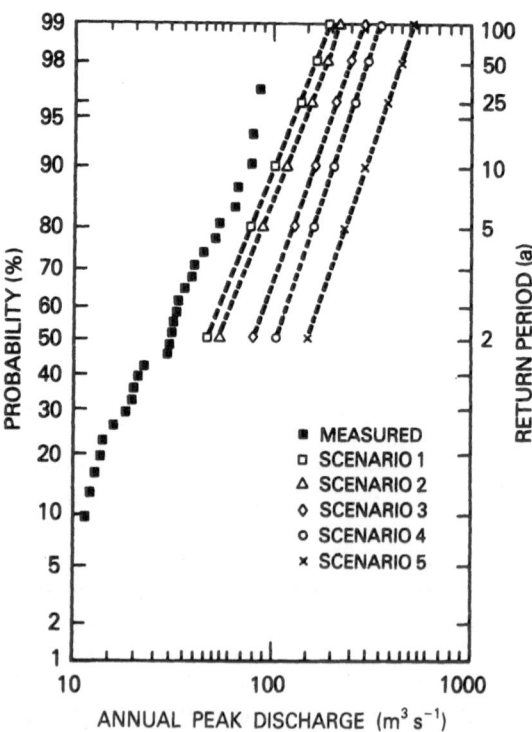

Figure 6.2 Comparison of the peak discharges for given return periods and five possible scenarios for the damage of a forest in a watershed (after Mauser 1984).

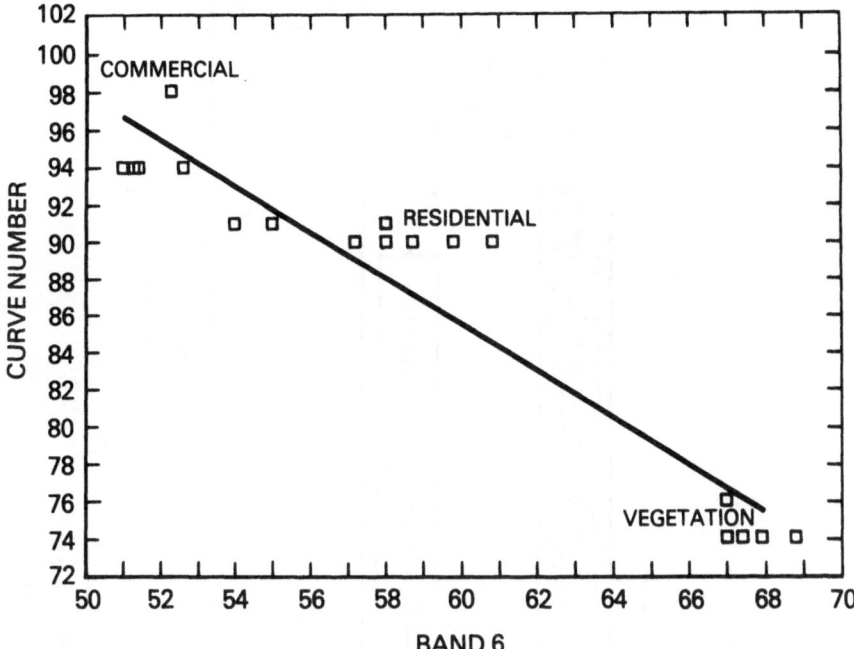

Figure 6.3 An illustration of the average reflectance for Landsat band 6 plotted against curve number (after McGregor 1987).

empirically related to the basin curve number. This concept is illustrated in Figure 6.3 which is taken from work by McGregor (1987) and his study of an urbanizing basin in Texas. A somewhat similar study by Zevenbergen *et al.* (1988) found that several reflectance models could be used to give good estimates of rangeland curve numbers.

6.5 NEW MODELS

It is apparent that some improvement in forecasting and simulation can be achieved by modifying existing models to use remote sensing data. However, it follows that even greater gains can be achieved with models designed to use remote sensing as well as conventional data. Such models would resemble contemporary simulation models structurally but would be able to account for the spatial variability found in natural basins in a more realistic way. Also, the subprocess algorithms (i.e. infiltration, evapotranspiration, etc.) would be written to use remote sensing data as a primary input as well as the more typical inputs.

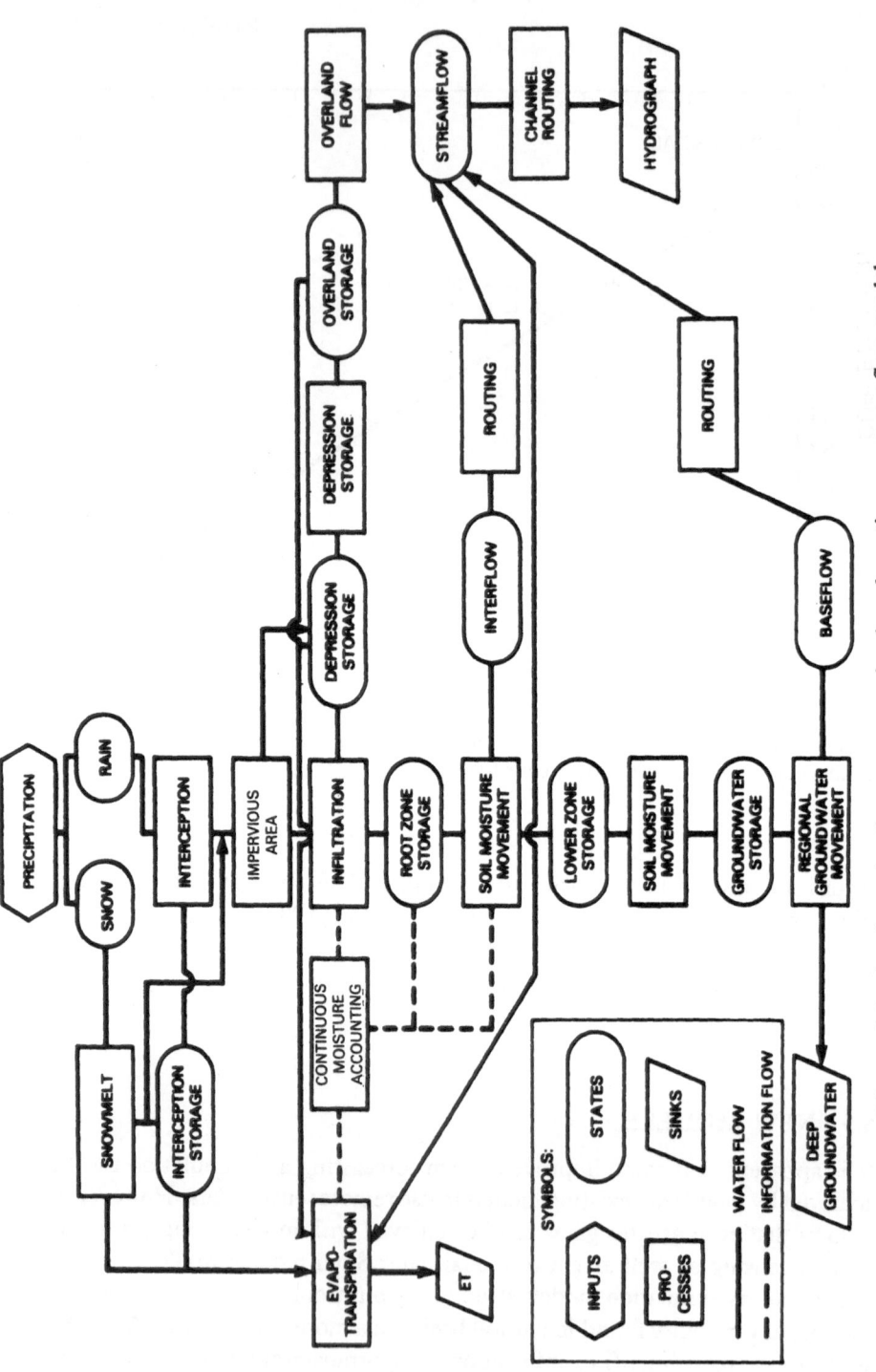

Figure 6.4 A schematic of a remote-sensing-based continuous streamflow model (after Groves and Ragan 1983).

A pioneering attempt to develop a model designed to use remote sensing input data has been made by Groves and Ragan (1983). This model is similar in structure to other models (Figure 6.4) such as the Stanford model (Crawford and Linsley 1966), but more of its parameters are physically based in the sense that they can be determined directly by remote sensing. Another feature of this model is the use of a geographic information system (GIS) as a data management tool to handle the spatial nature of the various data. The GIS assimilates remote sensing data and the historically more common point data and provides a spatially distributed framework for the model (Figure 6.5).

The US Army Corps of Engineers (1987) developed a microcomputer-based system that combines remote sensing image processing and spatial data analysis through a GIS. The system provides options to use the SCS curve number or the Snyder unit hydrograph to estimate runoff from single storm events. Remote sensing can be incorporated into the system in a variety of ways: as measures of land use, impervious surfaces, providing initial conditions for flood forecasting, and monitoring flooded areas.

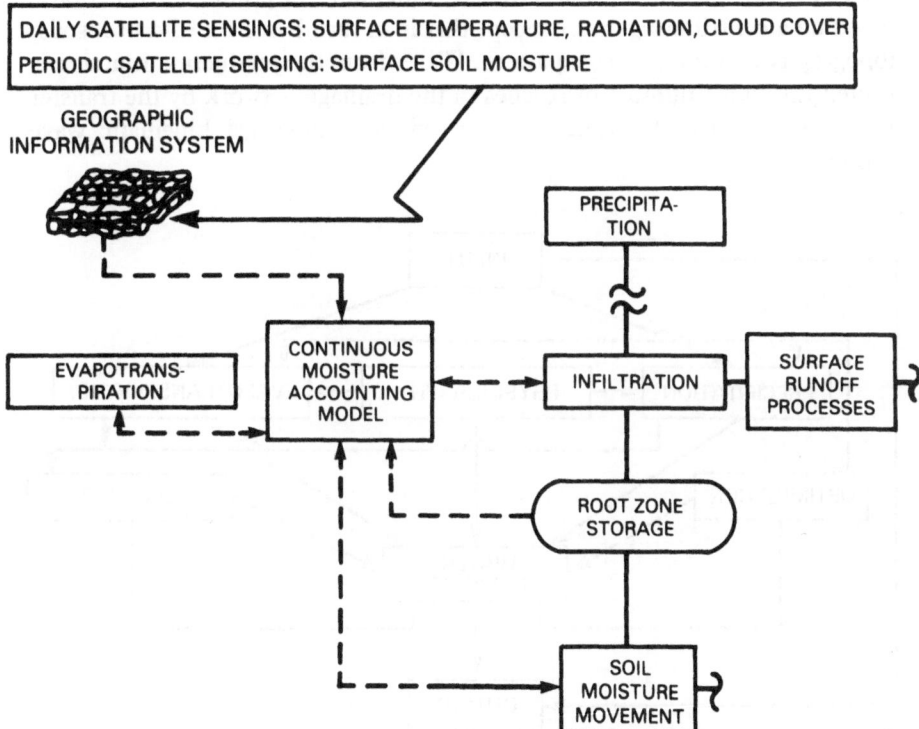

Figure 6.5 A schematic illustrating the information flow for the continuous moisture profile accounting procedure (after Groves and Ragan 1983).

Several investigators have proposed the use of remotely sensed rainfall as inputs to real-time hydrology and short-time forecast models. Rott (1986) has studied the potential for using Meteosat data as an indication of rainfall input to a basin. A runoff model that uses a cloud index developed from the thermal data was applied to two large basins in Europe. The concept was demonstrated by predicting runoff for 1 to 3 days ahead, starting with the measured flow.

Schultz (1986) has proposed using historic NOAA satellite data to generate a long-term runoff record for data-sparse basins. A cloud cover index developed from cloud-top temperatures is transformed into runoff with a linear transformation function. For short-term flood forecast problems, Schultz has also proposed developing rainfall inputs from near-real-time satellite data, and, if available, combining them with ground-based radar data. The flood flows are then calculated with a runoff model capable of using this type of precipitation input.

A somewhat analogous effort has been proposed by Fortin *et al.* (1986) that has nine modules (Figure 6.6). The framework of the model is built on a square grid information system with variable-sized pixels to incorporate many types of remote sensing data. The hydrology module is really two modules (Figure 6.7), a production function and a transfer function. The production function is based on a contributing area concept that is in turn based on land use, soils, topography and the drainage pattern. Water from the production function is routed through a number of reaches in the drainage network by the transfer function. The transfer function is based on a modified kinematic wave formulation.

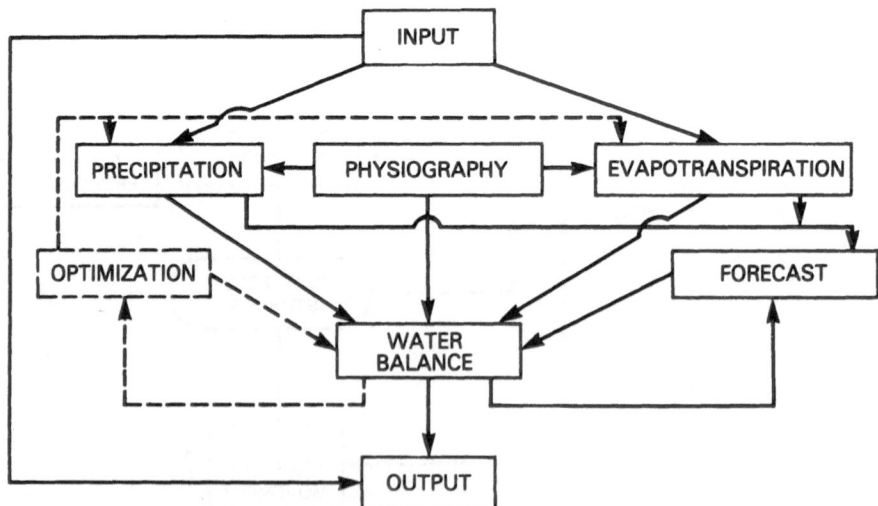

Figure 6.6   A schematic of a model structure for using remote sensing data at the input module (after Fortin *et al.* 1986).

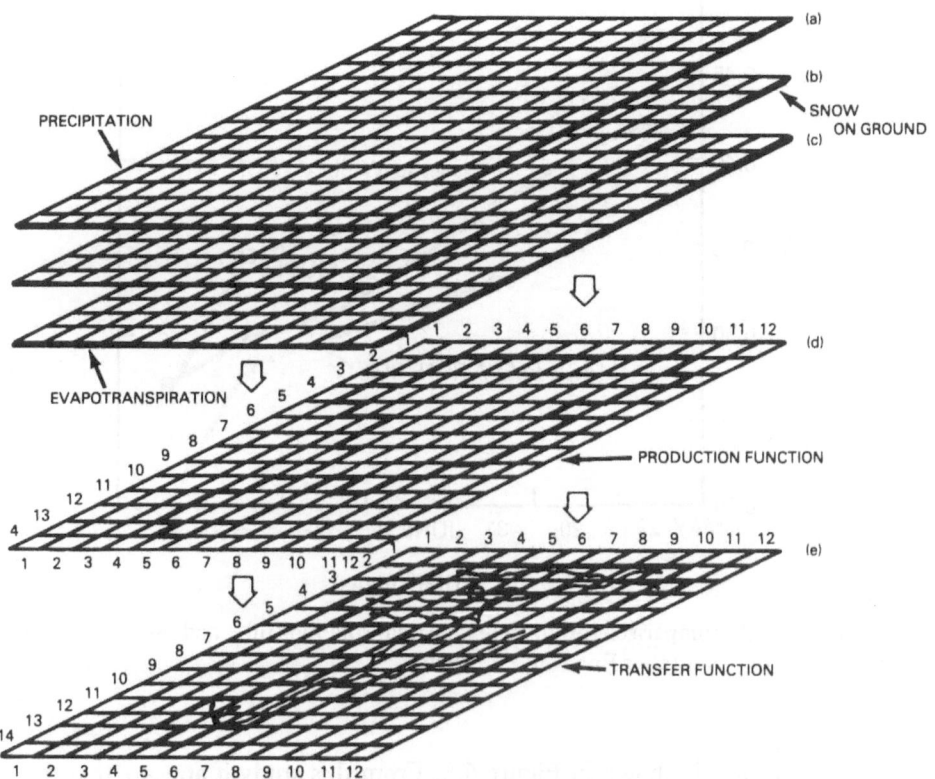

Figure 6.7 Spatial structure of model showing various hydrologic processes and production and transfer functions (after Fortin *et al*. 1986).

Kite (1989) has developed a database system for incorporating ground-based meteorological and hydrologic data and remote sensing data for use by a hydrologic model. To date, he has been evaluating the utility of Channels 1 and 2 (visible and near infrared) from the NOAA AVHRR for snow cover and cloud information. He has also used Landsat for establishing the land cover classification.

Engman *et al*. (1989) have used the time series of soil moisture maps (see Chapter 5) as the basis for simulating the water balance of a small basin in Kansas with a simple storage model. Aircraft microwave measurements of soil moisture were used to construct two-dimensional maps of the spatial distribution of the soil moisture. The use of data from flights on different dates also provided the temporal changes resulting from soil drainage and evapotranspiration. A comparison of the model-simulated soil moisture and

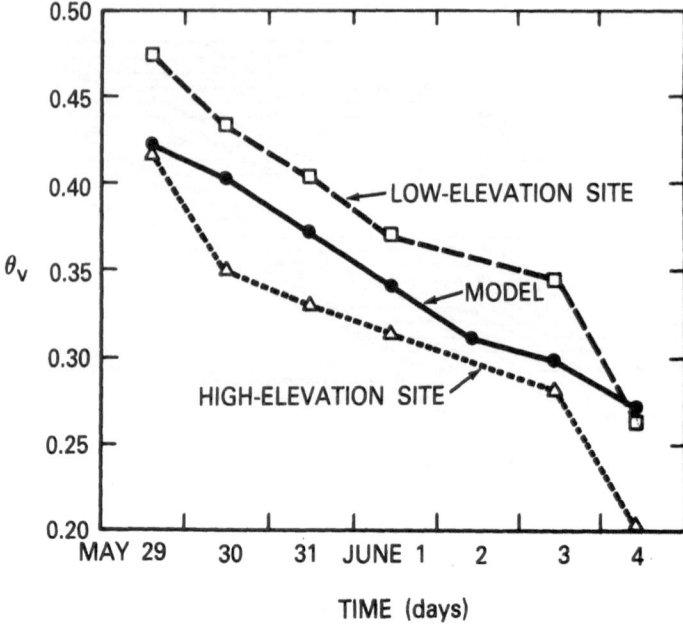

Figure 6.8 A comparison of model-simulated soil moisture and measured soil moisture at two sites (after Engman *et al.* 1989).

the ground data is shown in Figure 6.8. From this study it appears that the remotely sensed soil moisture will be a valuable new data form for verifying model performance and identifying areas of the basin contributing to streamflow.

Additional research on spatial variability of hydrologic parameters should provide rules for extrapolating point data. These rules or algorithms could be built into the GIS to use data from any source or location in the basin.

6.6 FUTURE APPLICATIONS

A number of hydrologists have speculated on how remote sensing will have an impact on hydrology and help to improve our data and models. A general consensus is that remotely sensed data will be extremely valuable but current techniques must be modified or new procedures developed to use these new data forms optimally. Salomonson *et al.* (1975) studied remote sensing requirements based on a sensitivity analysis of input parameters of the Kentucky watershed model (James 1970; Striffler 1973), a modification of the Stanford IV watershed model (Crawford and Linsley 1966). Of 26 input parameters, six could be determined by remote sensing. The results of a case study demonstrated that more current data from Landsat could improve runoff simulation. The authors cautioned about extending these results to other areas or other models. They varied only one parameter at a time but acknowledge that in reality the parameters are not likely to be independent. In spite of the caution, this study did show some of the potential that may be expected from remote sensing.

Others have studied remote sensing applications to hydrologic simulation models. Link (1983) provided a thorough coverage of the state of modelling and the use of remote sensing data. Engman (1982) reviewed the then current applications of remote sensing to hydrology which were relatively straightforward extensions of photogrammetry. He pointed out, however, that in the future the use of microwave and thermal data may provide very unique forms of information that may dramatically improve our ability to simulate, forecast and understand hydrologic events in that they may provide direct measurements of hydrologic state variables.

Peck *et al.* (1981) conducted a detailed study on the suitability of seven hydrologic forecasting and simulation models to assimilate remotely sensed data. In general, they concluded that remote sensing had limited usefulness for those models in their form at that time. Their study identified model variables in addition to land use and snow cover area that could be provided by remote sensing. These included soil moisture, frozen soils and snow water equivalent.

These three characteristics are currently not used as input data to any of the models discussed here. However, because of their importance in determining runoff rates and volumes, it appears that demonstrable improvements in forecast accuracy may be achieved if these quantities could be measured for input data. For example, Wiesnet (1976) stated that hydrologists in the NOAA River Forecast Center in Kansas City, Kansas, believed soil moisture to be the most troublesome parameter affecting their forecasts.

Improved forecasting depends upon two major factors: more physically realistic models to simulate the hydrologic process and adequate data to drive these models. Forecast improvement requires parallel developments in both of these areas. Several studies have demonstrated the importance of adequate data in both spatial coverage and type. Parmele *et al.* (1972) illustrated the importance of dense rain gauge networks to provide adequate input data. They used the Stanford model (Crawford and Linsley 1966) to show how forecast accuracy decreases as the gauge density for estimating input precipitation decreases.

Morton (1982) presented the argument that the commonly used simulation models require assumptions that cannot be supported by theory or empirical studies. For the most part, he claimed, the simulated models stress mathematical tractability and attempt to make reality conform to conventional wisdom. He proposed an analytical approach based on spatial averages of the major water balance components. He suggested independent estimates of areal evapotranspiration because spatial estimates of changes in storage (soil moisture, wetlands, lakes, etc.) are too costly to contemplate using even for a small basin. This may not be the case with remote sensing. It appears likely that we will be able to measure changes in storage (soil moisture, snow, surface water bodies) on a spatial scale useful for hydrologic modelling.

For the purpose of illustration, the following discussion will focus on soil moisture as the unique type of information that remote sensing can provide for hydrologic models. Although soil moisture is the focus of the discussion, the same general ideas can be extended for any of the other remote sensing variables.

Jackson *et al.* (1981) demonstrated how possible applications of repetitive, remotely sensed soil moisture might be used. They discussed how areal data may be used to calibrate soil and vegetation parameters and to correct errors resulting from point measurements of precipitation. In the study, they demonstrated how soil moisture observations are useful in calibrating and updating the state of the system. However, it was also pointed out that the model structure itself may preclude a valid analysis of the value of soil moisture measurements or of the frequency needed to improve simulations.

In general, existing models represent the soil in a way to make the model work and their designers have not considered the possibility of independent measures of soil moisture or soil parameters. This also reinforces Morton's (1982) view that our models primarily emphasize getting answers at the expense of modelling faithfully the physical processes involved. However, this approach was justified because soil moisture data have not been available and hydrologists have not been able to account for the spatial variability of soils in natural basins. In these models, soil moisture is a system state that must be initialized and then recomputed at successive times, being increased to account for added rainfall and decreased to reflect evapotranspiration and drainage. The level of soil moisture controls the rate of these as well as the rate of infiltration which is very sensitive to initial soil moisture content.

Since existing conceptual hydrologic models have been designed and calibrated for point measurements and allow soil moisture to be a fitting parameter, actual measures of the 'true' areal value of soil moisture has been of no value. The state of moisture conditions conceptually indicated by existing models probably does not relate to the actual field conditions. Thus, with present models, improved measurement of actual conditions may result in very little improvement in the accuracy or timeliness of hydrologic forecasts or simulations. In order to use measured soil moisture for forecasting, the models will have to be modified or new models developed to use remotely measured soil moisture as input data or as feedback to check on predictions and to update system states.

The possibility of using soil moisture as input data would allow the hydrologists to redesign the models so that the soil zone, subsurface flow to groundwater, interflow or base flow can be conceptualized in a more physically realistic manner. The surface runoff could be simulated by an infiltration model whose parameters could be either determined independently or calibrated. Likewise, the subsurface water transport could be modelled using procedures based on our knowledge of flow in porous media.

Soil moisture could also be used in a feedback loop to update the model state. Most models are based on a mass balance, taking rainfall (or snowmelt) as input and routing it to streamflow. Usually a portion is stored temporarily as soil moisture. The soil moisture defines the state of the system and, as such, controls the rate of sequential processes (i.e. infiltration or deep seepage) and events. If there are errors in the initial state of the model, errors in the predicted output may increase with time, since each successive computation is based on the previous state of the system. The prediction accuracy of models may be improved significantly if the system state could be checked periodically and updated or corrected as necessary to reduce the propagation of initialization errors. Repetitive measures of soil moisture used as feedback to the model could function in this way. This is one of the many possible future applications of remote sensing to hydrology.

Remote sensing is beginning to have an important role in determining runoff from rainfall. The remote sensing product can take the place of conventional maps and for many parts of the world these products may be the only source of map types of information. However, the real impact of remote sensing will not be in providing up-to-date maps to hydrologists, but rather in providing new types of data on the state of the hydrologic system. The work to date on determining land use is an example of how remote sensing can be used. In this case, reflectance values have been interpreted into land use classes that, when adapted properly, have been shown to be cost-effective and accurate inputs to land-use-based runoff models. In the future, thermal infrared and microwave measurements should prove to be even more valuable. Measurements in these parts of the electromagnetic spectrum show the potential to determine such basic watershed state variables as soil moisture, frozen soils and basic energy balance parameters. These may eventually be shown to be basic input data much like the soil data and land use data we use today.

REFERENCES

Allord, G. J. and Scarpace, F. L. (1979) Improving streamflow estimates through use of Landsat. In *Satellite Hydrology, 5th Annu. William T. Pecora Memorial Symp. on Remote Sensing*, American Water Resources Association, Minneapolis, MN, pp. 284–91.

Bondelid, T. R., Jackson, T. J. and McCuen, R. H. (1982) Estimating runoff curve numbers using remote sensing data. *Proc. Int. Symp. on Rainfall–Runoff Modeling*.

Applied Modeling in Catchment Hydrology, Water Resources Publications, Littleton, CO, pp. 519–28.

Chandra, S. and Sharma, K. P. (1978) Applications of remote sensing to hydrology. *Proc. Symp. on Hydrology of Rivers with Small and Medium Catchments, Roorkee, India*, vol. II, pp. 1–13.

Crawford, N. H. and Linsley, R. K. (1966) Digital simulation in hydrology: Stanford watershed model IV. *Stanford University Tech. Rep. 39*.

Draper, S. E. and Rao, S. G. (1986) Runoff prediction using remote sensing imagery. *Water Resour. Bull.* 22, 941–9.

Engman, E. T. (1982) Remote sensing application in watershed modeling. *Proc. Int. Symp. on Rainfall–Runoff Modeling. Applied Modeling in Catchment Hydrology*, Water Resources Publications, Littleton, CO, pp. 473–94.

Engman, E. T., Angus, G. and Kustas, W. P. (1989) Relationships between the hydrologic balance of a small watershed and remotely sensed soil moisture. *Proc. IAHS 3rd Int. Assembly, Baltimore, MD, IAHS Publ. No. 186*, pp. 75–84.

Fortin, J. P., Villeneuve, J. P., Guillot, A., Sequin, B. (1986) Development of a modular hydrological forecasting model based on remotely sensed data, for interactive utilization on a microcomputer. *Hydrologic Applications of Space Technology, IAHS Publ. No. 160*, pp. 307–19.

France, M. J. and Hedges, P. D. (1986) A hydrological comparison of Landsat TM, Landsat MSS and black and white aerial photography. *Symp. on Remote Sensing for Resources Development and Environmental Management, Enschede, The Netherlands*, pp. 717–20.

Groves, J. R. and Ragan, R. M. (1983) Development of a remote sensing based continuous streamflow model. *Proc. 17th Int. Symp. on Remote Sensing of Environment, Ann Arbor, MI*, Environmental Research Institute of Michigan, Ann Arbor, MI, pp. 447–56.

Gugan, D. J. and Dowman, I. J. (1988) Topographic mapping from SPOT imagery. *Photogram. Eng. Remote Sensing* 54, 1409–14.

Haralick, R. M., Wang, S., Shapiro, L. G. and Campbell, J. B. (1985) Extraction of drainage networks by using a consistent labeling technique. *Remote Sensing Environ.* 18, 163–75.

Inglis, C. C. (1939). The relationship between meandering belts, distance between meanders on axis of stream, width and discharge of rivers in flood plains and incised rivers. Government of India, Central Board of Irrigation and Power, Annual Report.

Jackson, T. J. and Ragan, R. M. (1977) Value of Landsat in urban water resources planning. *ASCE J. Water Resour. Plann. Manage. Div.* 103, No. WR1, Proc. Paper 12906, pp. 33–46.

Jackson, T. J., Ragan, R. M. and Fitch, W. N. (1977) Test of Landsat-based urban hydrologic modeling. *ASCE J. Water Resour. Plann. Manage. Div.* 103, No. WR1, Proc. Paper 12950, pp. 141–58.

Jackson, T. J., Ragan, R. M. and Shubinski, R. P. (1976) Flood frequency studies on ungaged urban watersheds using remotely sensed data. *Proc. Natl. Symp. on Urban Hydrology, Hydraulics and Sediment Control, University of Kentucky, Lexington, KY*, University of Kentucky, pp. 31–9.

Jackson, T. J., Schmugge, T. J., Nicks, A. D., Coleman, G. A. and Engman, E. T.

(1981) Soil moisture updating and microwave remote sensing for hydrologic simulation. *Hydrol. Sci. Bull.* **26**, pp. 305–19.

James, L. D. (1970) An evaluation of the relationships between streamflow patterns and watershed characteristics through the use of OPSET — a self calibrating version of the Stanford watershed model. US Dept of Interior, Contract No. 14-01-0001-1964, University of Kentucky, Water Resources Institute, Lexington, KY.

Kite, G. (1989) Using NOAA data for hydrologic modeling. *Proc. IGARSS'89, Vancouver, BC, Canada,* vol. 2, pp. 553–7.

Link, L. E. (1983) Compatibility of present hydrologic models with remotely sensed data. *Proc. 17th Int. Symp. Remote Sensing of Environment, Ann Arbor, MI,* Environmental Research Institute of Michigan, Ann Arbor, MI, pp. 133–53.

McGregor, K. M. (1987) Using Landsat to derive curve numbers for hydrologic models. In *American Society for Photogrammetry and Remote Sensing and ASCM Fall Convention, Reno, NV, ASPRS Tech. Paper,* pp. 129–35.

Mauser, W. (1984) Calculation of flood hydrographs using satellite-derived land-use information in the Dreisam watershed/S-W Germany. *Proc. IGARSS'84 Symp.,* ESA-SP-215, pp. 301–4.

Morton, F. I. (1982) Integrated basin response — A problem of synthesis or a problem of analysis. *Proc. Canadian Hydrology Symp.,* Associated Committee on Hydrology, National Research Council, Canada, pp. 361–3.

Parmele, L. H., Engman, E. T. and Hendrick, R. L. (1972) The use of simulation models for evaluating the efficiency of a precipitation gage network. *Proc. 2nd Symp. on Meteorological Observations and Instrumentation,* American Meteorological Society, San Diego, CA, pp. 52–7.

Peck, E. L., Keefer, T. N. and Johnson, E. R. (1981) Strategies for using remotely sensed data in hydrologic models. *NASA Rep. No.* CR-66729, Goddard Space Flight Center, Greenbelt, MD.

Ragan, R. M. and Jackson, T. J. (1980) Runoff synthesis using Landsat and SCS model. *J. Hydraul. Div., ASCE* **106**, (HY5), 667–78.

Rango, A., Feldman, A., George, III, T. S. and Ragan, R. M. (1983) Effective use of Landsat data in hydrologic models. *Water Resour. Bull.* **19**, 165–74.

Rott, H. (1986) Satellite data as input for long-term and short-term hydrological models. *Hydrologic Applications of Space Technology, IAHS Publ. No.* 160, pp. 321–30.

Salomonson, V. V., Ambaruch, R., Rango, A. and Ormsby, J. P. (1975) Remote sensing requirements as suggested by watershed model sensitivity analysis. *Proc. 10th Symp. on Remote Sensing of Environment, Ann Arbor, MI,* Environmental Research Institute of Michigan, Ann Arbor, MI, pp. 1273–84.

Schultz, G. A. (1986) Satellite data as input for long-term and short-term hydrological models. *Hydrologic Applications of Space Technology, IAHS Publ. No.* 160, pp. 297–306.

Shope, Jr, W. G. and Paulson, R. W. (1986) Development of a national real-time hydrologic information system using GOES satellite technology. *Hydrologic Applications of Space Technology, IAHS Publ. No.* 160, pp. 13–21.

Slack, R. B. and Welch, R. (1980) Soil conservation service runoff curve number estimates from Landsat data. *Water Resour. Bull* **16**, 887–93.

Still, D. A. and Shih, S. F. (1985) Using Landsat data to classify land use for assessing the basinwide runoff index. *Water Resour. Bull.* **23**, 931–40.

Striffler, W. D. (1973) User's manual for Colorado State University version of the Kentucky watershed model. *Contract Rep., NASA Contract* NAS9-13142, Colorado State University, Fort Collins, CO.

Tholin, A. L. and Keifer, C. J. (1959) The hydrology of urban runoff. *ASCE J. Sanit. Eng. Div.* **85**, 47–106.

Thomas, D. M. and Benson, M. A. (1970) Generalized streamflow characteristics from drainage basin characteristics. *USGS Water Supply Pap.* 1955, Washington, DC.

Ungar, S. G., Merry, C. J., Irish, R., McKim, H. L., Miller, M. S. (1988) Extraction of topography from side-looking satellite systems — A case study with SPOT simulation data. *Remote Sensing Environ.* **26**, 51–73.

United Nations (1955) Economic Commission for Asia and the Far East. Multi-purpose river basin development, part 1, manual of river basin planning flood control series no. 7, United Nations Publication ST/ECAFE/SERF/7, New York.

US Army Corps of Engineers (1976) Urban storm water runoff, 'STORM'. Computer program 723-L2520, Hydrologic Engineering Center, Davis, CA.

US Army Corps of Engineers (1987) Remote sensing technologies and spatial data applications. *Res. Doc. No.* 29, pp. 1–5, Hydrologic Engineering Center, Davis, CA.

US Department of Agriculture (1969) Computer program for project formulation. *Tech. Release No.* 20, Soil Conservation Service, Washington, DC.

US Department of Agriculture (1972) Soil Conservation Service. *National Engineering Handbook*, section 4, *HYDROLOGY* US Govt Printing Office, Washington, DC.

US Department of Agriculture (1986) Urban hydrology for small watersheds. *Tech. Release No.* 55, 2nd edn, Soil Conservation, Washington, DC.

Wiesnet, D. R. (1976) Remote sensing and its application to hydrology. In *Facets of Hydrology* (ed. J. C. Rodda), Wiley, Chichester.

Zevenbergen, A. W., Rango, A., Ritchie, J. C., Engman, E. T. and Hawkins, R. H. (1988) Rangeland runoff curve numbers as determined from Landsat MSS data. *Int. J. Remote Sensing* **9**, 495–502.

7

Soil moisture

7.1 INTRODUCTION

Soil moisture is the temporary storage of precipitation within a shallow layer of the Earth that is generally limited to the zone of aeration, which approximately coincides with the root zone. Precipitation that is temporarily stored as soil moisture can follow several paths. It can be returned directly to the atmosphere through direct evaporation from the soil surface or by way of plants through transpiration, or it can percolate to a phreatic (saturated) zone as groundwater recharge and eventually be transmitted as groundwater flow to stream channels. The time period that water exists as soil moisture is usually relatively short, being of the order of hours and days. If it is frozen, the soil moisture residence time may be months or essentially permanent in the case of permafrost. Soil moisture is highly variable as the result of the inhomogeneity of soil properties, topography, land cover and the non-uniformity of input from rainfall. Traditionally soil moisture has been difficult to measure in a way that is representative of anything except a point. Typically, averages of point measurements are used to characterize the soil moisture of an area, but these averages seldom yield information that is representative to characterize other hydrologic processes (i.e. evapotranspiration, runoff, groundwater recharge, etc.).

There is a need to develop ways to quantify the spatial variability of soil moisture; to determine the scale of hydrologic units that characterize various hydrologic processes; and to develop criteria for delineating areas that can be treated as hydrologically uniform. Remotely sensed data have great potential for providing areal estimates of soil moisture rather than point measurements.

Because hydrologic concepts have been developed from point measurements (i.e. rain gauges, soil columns, etc.), hydrologists have been generally unsuccessful in treating spatial variability.

There is also a value in acquiring continental soil moisture maps on a frequent basis to isolate regions with strong soil moisture gradients in both time and space. This type of spatially derived information could be used to monitor the effectiveness of precipitation as an input to a region and to provide information for meteorological and global circulation models.

7.2   GENERAL APPROACH

Remote sensing of soil moisture can be accomplished to some degree or other by all regions of the electromagnetic spectrum. Successful measurement of soil moisture by remote sensing techniques depends upon the type of reflected or emitted radiation. Table 7.1 summarizes the advantages and disadvantages of each approach.

A comprehensive summary of remote sensing approaches for measuring soil moisture has been presented by Schmugge *et al.* (1980). However, as will be discussed in the following subsections, only the microwave region offers the potential for truly quantitative measurements from space.

Gamma radiation techniques

Airborne soil moisture measurement by gamma radiation is based on detecting the difference between the natural terrestrial gamma radiation flux for wet and dry soils. The presence of water in the upper soil layers increases the attenuation of the gamma radiation from below; thus the flux is less for wet soils than for dry soils. Quantitative estimates of soil moisture require calibration flight lines to determine a background soil moisture value, M_0, and a background gamma count rate, C_0. Then the current soil moisture, M can be calculated according to

$$M = \frac{C/C_0(100 + 1.11M_0) - 100}{1.11} \qquad (7.1)$$

where C is the measured gamma count rate. A more complete description can be found in Carroll (1981). Because the atmosphere also attenuates the gamma radiation flux from the soil, this approach is limited to aircraft flying at an altitude of only 100–200 m above the Earth's surface.

Visible/near-infrared techniques

Reflected solar radiation is not a particularly viable technique for measuring soil moisture because there are too many noise elements that confuse the

Table 7.1 Summary of remote sensing techniques for measuring soil moisture (Engman 1982)

Wavelength region	Property observed	Advantages	Disadvantages or noise sources
Reflected solar	Albedo; index of reflection	Data available	No unique relationship between spectral reflectance and soil moisture; thin surface layer only; cloud interference
Thermal infrared	Surface temperature (measured diurnal range of surface temperature or crop canopy temperature	High spatial resolution, large swath; relationship between temperature and soil water pressure is independent of soil type	Bare soil only; cloud interference; surface topography and local meteorological conditions can cause noise; surface layer only (2–4 cm)
Active microwave (1–100 cm)	Backscatter coefficient; dielectric constant	All-weather high resolution; limited swath width	Surface roughness; vegetation; topography
Passive microwave (1–100 cm)	Brightness temperature (microwave emission); dielectric constant; soil temperature	All-weather; penetrates some vegetation; large area coverage	Limited spatial resolution; soil temperature; surface roughness; vegetation; interference from communications

interpretation of the data. Although wet soil will generally have a lower albedo than dry soil (Crist and Cicone 1984), and this difference can be measured theoretically, confusion factors such as organic matter, roughness, texture, angle of incidence, colour, plant cover and the fact that it is a transient phenomenon all make this approach impractical (Jackson *et al.* 1978).

Thermal techniques

After meteorological inputs to the soil surface have been accounted for, surface temperature is primarily dependent upon the thermal inertia of the soil. The thermal inertia, in turn, is dependent upon both the thermal conductivity and the heat capacity which increases with soil moisture according to the relationship (Price 1982)

$$DT_s = T_s\,(\text{PM}) - T_s\,(\text{AM}) = f(1/D) \qquad (7.2)$$

where DT_s is the diurnal temperature difference between the afternoon surface temperature (T_s (PM)) and the early morning temperature (T_s (AM)), and D is the diurnal thermal inertia given by

$$D = \omega \varrho_c\, k \qquad (7.3)$$

where ω corresponds to the day length, ϱ_c is the volumetric heat capacity and k is the thermal conductivity. The diurnal thermal inertia D describes the ability of the soil to resist temperature change. For example, a dry sand has a relatively low value of D compared with a wet clay because the thermal conductivity of sand is lower than that for clay and the volumetric heat capacity value for dry soil is lower than that for wet soils. Thus, by measuring the amplitude of the diurnal temperature change, we can develop a relationship between the temperature change and soil moisture. However, the relationship between diurnal temperature and soil moisture depends upon soil type and is largely limited to bare soil conditions (van de Griend *et al.* 1985).

Microwave techniques

Microwave techniques for measuring soil moisture include both the passive and active microwave approaches with each having distinct advantages. The theoretical basis for measuring soil moisture by microwave techniques is based on the large contrast between the dielectric properties of liquid water and dry soil. The large dielectric constant for water is the result of the water molecule's alignment of the electric dipole in response to an applied electromagnetic field. The dielectric constant of water is approximately 80 compared with that of dry soils which is of the order of 3 to 5. Thus, as the soil moisture increases, the dielectric constant can increase to a value of 20 or more (Schmugge 1983). Figure 7.1 illustrates the change in dielectric constant for soil at several microwave frequencies. For passive microwave remote sensing, this change in dielectric constant would result in a decrease of emissivity from about 0.95 to 0.6 or less. For the active case, the measured radar backscatter would increase by about 10 dB, or possibly by even more.

Soil texture affects the microwave sensing of soil moisture in the way that the dielectric constant changes with the relative amounts of sand, silt and clay in the soil. According to Schmugge (1983), the effect of soil texture can be understood by examining what happens to the water molecules as water is added to soil.

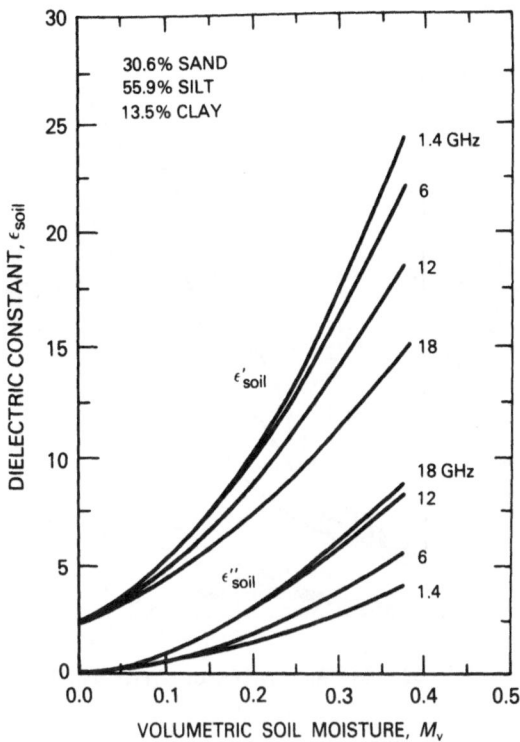

Figure 7.1   An illustration of the real and imaginary parts of the dielectric constant as a function of volumetric moisture content for a loamy soil measured at four frequencies (after Ulaby *et al.* 1986).

The large dielectric constant of liquid water is due to the molecule's ability to align its dipole moment along an applied field; thus anything that would hinder the molecule rotation, e.g. freezing, very high frequencies, or tight binding to a soil particle will reduce the dielectric constant of the water. Since the first water molecules which are added to the soil are tightly bound to the particle's surface, they will contribute only a small increase to the soil's dielectric constant. As more water is added, above some transitional level W_t, the additional molecules are farther away from the particle surface and are freer to rotate and make a larger contribution to the soil's dielectric constant. Since the surface area in a soil depends upon its particle-size distribution or texture, clay soils, with a larger surface area, will be able to hold more of this tightly bound water than sandy soils; thus, the transition point occurs at higher moisture levels in clay than in sandy soils.

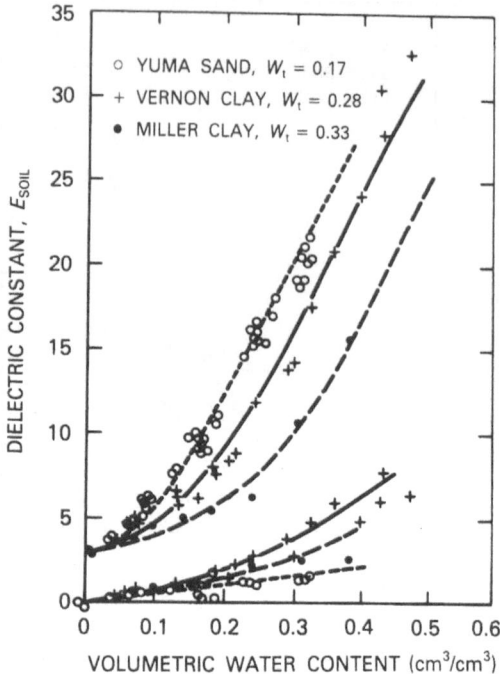

Figure 7.2 A comparison of laboratory measurements of the real and imaginary parts of the dielectric constant and model predictions (smooth curves) for three soils as functions of moisture content at a wavelength of 21 cm (after Wang and Schmugge 1980).

Figure 7.2 shows this effect with laboratory data and an empirical model developed by Wang and Schmugge (1980).

For passive microwave remote sensing of soil moisture, a radiometer measures the intensity of emission from the soil surface. This emission is proportional to the product of the surface temperature and the surface emissivity which is commonly referred to as the microwave brightness temperature (T_B), and can be expressed as follows:

$$T_B = t(H)[rT_{sky} + (1 - r)T_{soil}] + T_{atm} \qquad (7.4)$$

where $t(H)$ is the atmospheric transmission, r is the surface reflectivity, and T_{sky}, T_{soil} and T_{atm} are the temperatures for the sky, soil and atmosphere, respectively. At longer microwave wavelengths (greater than about 5 cm), which are better for measuring soil moisture, the atmospheric effects can be neglected and equation (7.4) can be simplified to

$$T_B = (1- r)T_{soil} = eT_{soil} \qquad (7.5)$$

where $e = 1 - r$ is the emissivity. Because the measured brightness temperature, and thus emissivity, is dependent upon soil texture, surface roughness and any vegetation present, actual soil moisture is usually related to T_B empirically with ground data.

For the active microwave approach, the measured radar backscatter, S_t, is made up of backscatter from vegetation, S_v, and soil, S_s, and the attenuation caused by the vegetation canopy, L. These terms can be related by the simple model

$$S_t = S_v + S_s/L \qquad (7.6)$$

The soil backscatter, S_s can be related directly to soil moisture by

$$S_s = R \; \alpha \; M_v \qquad (7.7)$$

where R is a surface roughness term, α is a soil moisture sensitivity term and M_v is the volumetric soil moisture. Although R and α are known to vary with wavelength, polarization and incidence angle, there is no satisfactory theoretical model suitable for estimating these terms independently. Thus, as is the case for the passive microwave approach, the relationship between measured backscatter and soil moisture requires an empirical relationship with ground data, even for bare soils.

An additional approach for using soil moisture data derived from microwave approaches is through change detection. This approach can be used for both passive or active microwave data. The change detection method minimizes the impact of target variables such as soil texture, roughness and vegetation because these tend to change slowly, if at all, with time. With change detection it is assumed that the only target change occurring is the soil moisture. Thus, any measured changes in T_B or S_t can be related directly to changes in soil moisture. Fortunately, both the brightness temperature and backscatter relationships with soil moisture are approximately linear. Also, fortunately for many hydrologic applications, the changes in soil moisture are more important than the actual absolute value of soil moisture.

7.3 CURRENT APPLICATIONS

Although not a truly operational technology yet, remote sensing of soil moisture has been shown to be possible through a large number of research projects. A great deal has been learned from these studies but additional work is needed to define the optimal sensor configurations. In addition, in contrast to some of the other aspects of hydrology, there is no existing satellite system that is collecting data suitable for soil moisture, other than perhaps thermal data. The current state of the science is reviewed in the following pages.

Thermal remote sensing of soil moisture

Thermal infrared measurements have been successfully used to measure the surface few centimetres of soil moisture. Idso *et al.* (1975) found that the volumetric moisture content for soil layers between 2 and 4 cm thick was linearly related to the amplitude of the diurnal soil temperature. Figure 7.3 shows the experimental results for four soil depths. However, tests on other soil types indicated that the results from one soil could not be accurately

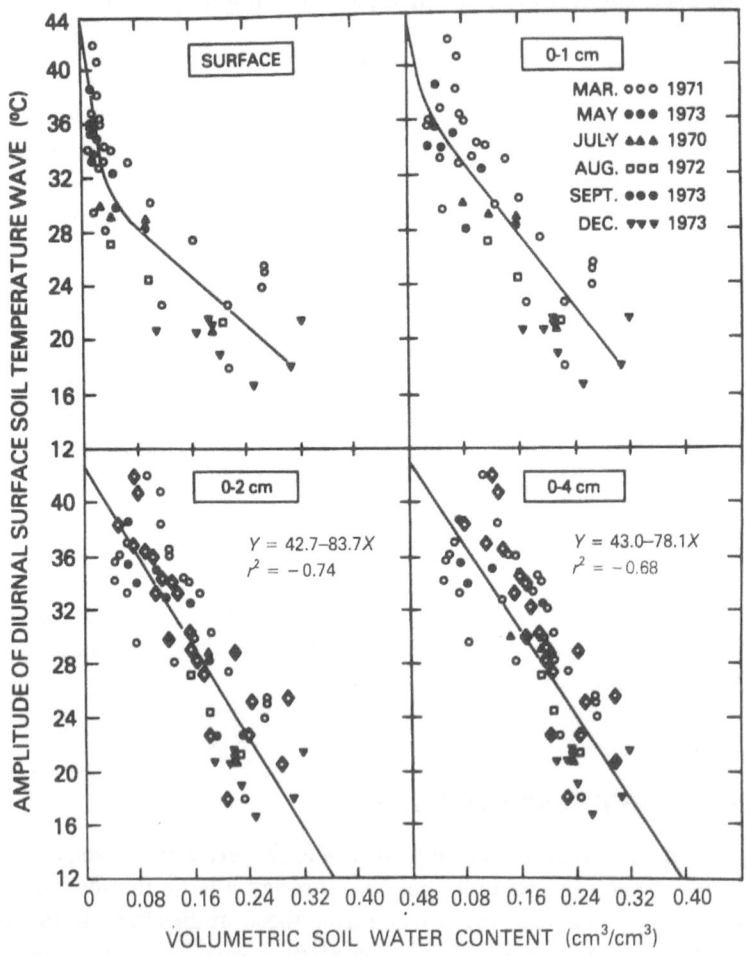

Figure 7.3 An illustration of the diurnal surface soil temperature variation against the mean daylight volumetric soil water content of four different depths (after Idso *et al.* 1975).

applied to other soils except when moisture was expressed as a pressure potential. By using the moisture characteristic curve for each soil type the moisture could be expressed as a pressure potential. One additional limitation to the thermal approach is that it cannot effectively be applied to surfaces with vegetation cover.

Van de Griend *et al.* (1985) not only demonstrated a relationship between night-time infrared data and soil moisture but also indicated that there was little sensitivity to soil moisture when a vegetation cover is present. Wetzel *et al.* (1984) used thermal data from geostationary satellites to demonstrate the feasibility of measuring soil moisture; however, no ground data were taken to verify the concept.

Moreover, in a study comparing aircraft thermal and soil moisture determined by C-band active microwave measurements, Perry and Carlson (1988) found little correlation between the two, especially in the spatial fields.

Microwave sensing of soil moisture

As discussed above, microwave techniques for measuring soil moisture have a strong theoretical basis. In addition, they are not limited to cloud-free and bare-soil conditions because the microwave approach can sense through cloud cover and, in many cases, through a vegetation canopy. Each of the two basic approaches, passive and active, offer different but distinct advantages. Each will be discussed below.

Passive microwave

In order to examine further the relationship between brightness temperature and soil moisture, it will be useful to re-examine equation (7.5). In this equation, the reflectance r is determined from the Fresnel equations for an electromagnetic wave at a smooth boundary. These can be written for the horizontally and vertically polarized radiation as follows:

$$r_H = \frac{K \cos \theta - \sqrt{K - \sin^2 \theta}}{K \cos \theta + \sqrt{K - \sin^2 \theta}} \qquad (7.8a)$$

$$r_V = \frac{\cos \theta - \sqrt{K - \sin^2 \theta}}{\cos \theta + \sqrt{K - \sin^2 \theta}} \qquad (7.8b)$$

where the subscripts H and V refer to the horizontal and vertical polarizations respectively, θ is the angle of incidence and K is the dielectric constant. Using a model like that proposed by Wang and Schmugge (1980), we can develop a relationship between soil moisture and emissivity as shown in Figure 7.4 and

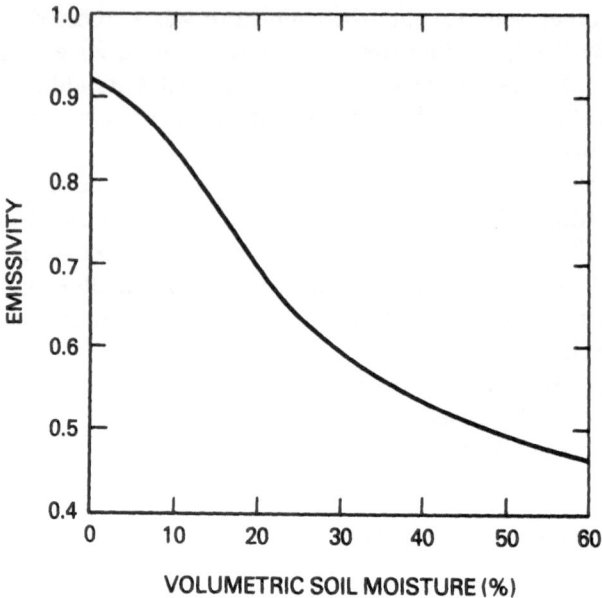

Figure 7.4 An illustration of the calculated emissivity against volumetric soil moisture for a smooth-surface sandy soil, $\theta° = 0$, $\lambda = 21$ cm, H polarization (after Jackson and Schmugge 1986).

brightness temperature as shown in Figure 7.5. These are the basic relationships for measuring soil moisture by passive microwave techniques and they have been verified by numerous examples of laboratory, truck and aircraft experiments as shown in Figures 7.6 and 7.7. An additional feature can be seen by comparing Figure 7.7(a) with 7.7(b). The increased sensitivity to soil moisture for the L-band as compared with the C-band data becomes clearly evident.

Although these relationships look reasonably good, there are several additional considerations that must be taken into account when measuring soil moisture by passive microwaves. These include the depth of measurement, surface roughness and vegetation effects. Each of these are discussed in more detail below:

(a) Measurement depth. The relationship between emissivity and soil moisture depends upon the dielectric constant across the air–soil interface. Consequently, this results in some uncertainty as to exactly how thick the soil layer is for determining the dielectric constant. According to Wilheit (1978), the layer of soil would be of the order of a few tenths of a wavelength. However, Mo *et al.* (1980) determined that the radiometric sampling depth is between 0.06 and 0.1 times the wavelength. In an experiment comparing

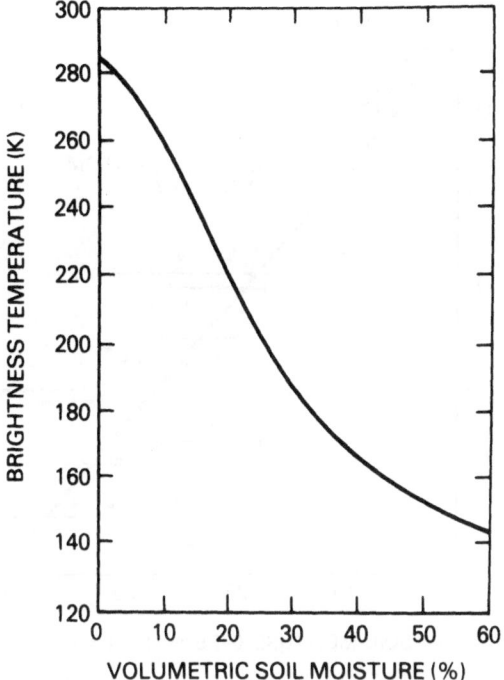

Figure 7.5 An illustration of the calculated brightness temperature against volumetric soil moisture for a smooth-surface sandy soil, $\theta = 0°$, $\lambda = 21$ cm, H polarization, $T_{eff} = 310$ K (after Jackson and Schmugge 1986).

drying measurements of soil layers at three frequencies, Newton *et al.* (1982) found that for L-band (21 cm wavelength) the sampling depth was about two-tenths of the wavelength. The measurement depth is not a constant but is related to the total amount of water in the soil layer, and thus the moisture content, and to the measurement frequency, as shown in Figure 7.8.

(b) Surface roughness. Microwave emission from the soil is related to the reflectivity of the surface, which if smooth can be calculated by the Fresnel equations (equations 7.8(a) and 7.8(b)). Smoothness in microwave terms is a relative term, being dependent upon the wavelength. That is, a surface that is smooth for one wavelength, say 21 cm L-band (1.4 GHz), may not be smooth for 6 cm C-band (4.9 GHz) and 2.8 cm K-band (10.7 GHz). The effect of a rough surface is to increase the surface emissivity and thus to decrease the sensitivity to soil moisture (Newton and Rouse 1980).

Choudhury *et al.* (1979) have shown that surface roughness can affect the soil reflectivity, R, in the following way:

$$R = R_0 \exp(-H \cos \theta) \tag{7.9}$$

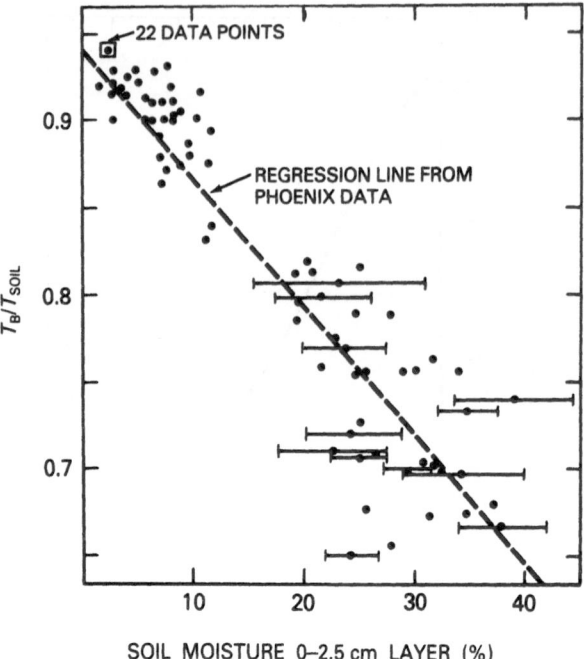

Figure 7.6 An illustration of aircraft observations of microwave brightness temperature against soil moisture in the top 2.5 cm during 1976 and 1977 flights over agricultural fields in Hand County, South Dakota. The horizontal error bars represent field sampling variability (after Schmugge *et al.* 1978).

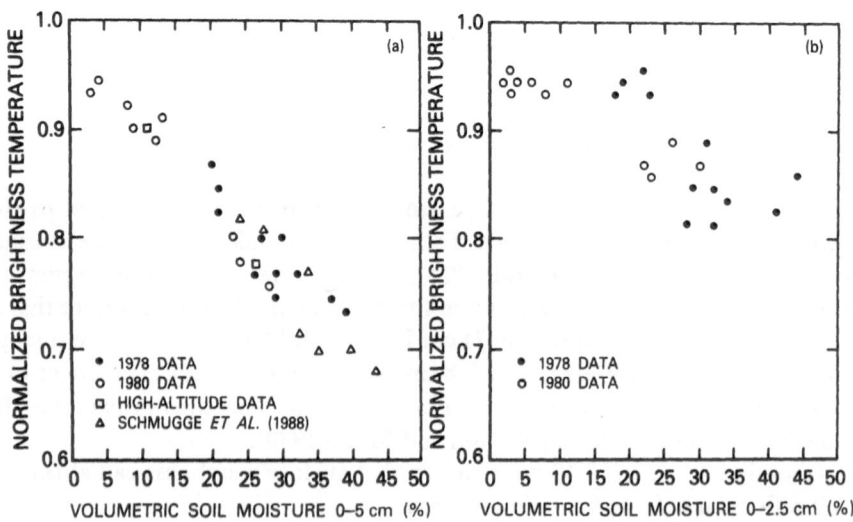

Figure 7.7 Aircraft emissivity and soil moisture observations over rangeland watersheds: (a) L-band, $\theta = 0°$, H polarization; (b) C-band, $\theta = 0°$, H polarization (from Jackson and Schmugge 1989).

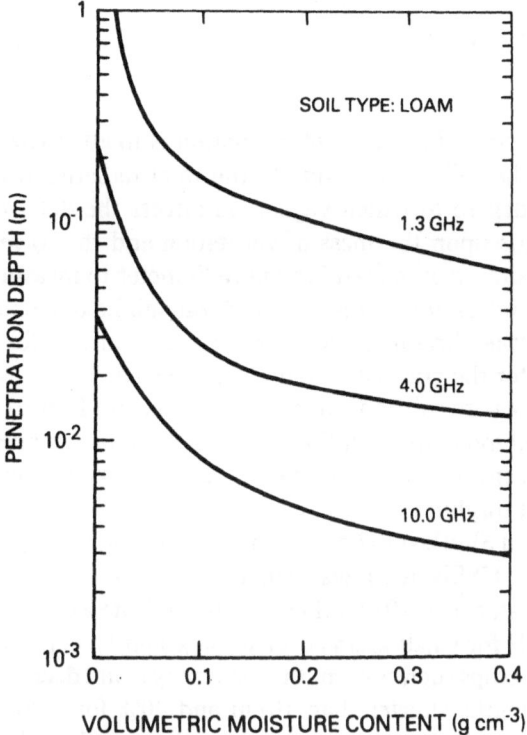

Figure 7.8 An illustration of penetration depth as a function of moisture content for three frequencies (after Ulaby *et al.* 1982).

where R_0 is the smooth-surface reflectivity, h is a roughness parameter proportional to the root mean square height variations of the soil surface, and θ is the angle of incidence. Wang *et al.* (1980) assumed that the random roughness was independent of the angle of incidence and simplified equation (7.9) to

$$R = R_0 \exp{(-h)} \tag{7.10}$$

More recent work by Promes *et al.* (1988) has shown that, for L-band, assuming smooth-field emissivity for most dry-land agricultural conditions in which the row height to row spacing is less than two will result in an error of less than 3%.

Theis *et al.* (1986) have demonstrated the possibility of using a multisensor approach for improving the estimates of soil moisture under field conditions. In this case, the effects of surface roughness were accounted for with scatterometer measurements. These were then used in a soil moisture equation which included terms related to the emissivity measured by the radiometer and to the scatterometer roughness term. Inclusion of the roughness term improved the R^2 values from 0.22 to 0.65 for C-band and from 0.69 to 0.95 for L-band.

(c) Vegetation cover.   The effect of vegetation is to attenuate the microwave emission from the soil; it also adds to the total radiative flux with its own emission. The degree to which vegetation affects the determination of soil moisture depends upon the mass of vegetation and the wavelength. Barton (1978) used an aircraft-mounted 2.8 cm radiometer to measure soil moisture over bare soils and uniform grass cover. Although he demonstrated a strong relationship between brightness temperature and moisture for the bare fields, no relationship for the grass sites could be perceived.

In studies over bare soil and sorghum, Newton and Rouse (1980) found no sensitivity to soil moisture with the 2.8 cm measurements over the sorghum, but with the 21 cm data the radiometer was sensitive to soil moisture even under the tallest sorghum.

Basharinov and Shutko (1975) and Kirdiashev *et al.* (1979) studied a variety of crops in the USSR with wavelengths varying from 3 to 30 cm. For wavelengths greater than 10 cm, their results indicate a decrease in sensitivity of about 10–20% for small grains over what would be expected for bare soil. With broad-leaf crops such as corn, the sensitivity could decrease by as much as 80% for wavelengths shorter than 10 cm and 40% for a 30 cm wavelength. Thus, the wavelength effect can be seen from these studies, that is a vegetation canopy is more transparent to longer wavelengths than to shorter wavelengths.

Jackson *et al.* (1982) developed a parametric approach based on a theoretical model proposed by Basharinov and Shutko (1975). This model treats the vegetation as an absorbing layer that can be quantified in terms of the water content of the vegetation by the relationship

$$M_v = 78.9 - 78.4\,[1 + (e - 1)\,\exp\,(0.22W)] \qquad (7.11)$$

where M_v is the volumetric soil moisture (0–2.5 cm), e is the measured emissivity and W is the water content of the vegetation (kg m^{-2}). Figure 7.9 illustrates the effect of vegetation on soil moisuture. An additional advantage to the correction proposed by Jackson *et al.* (1982) is that all data needed in equation (7.11) can be measured with remote sensing.

Dead vegetation can also have an attenuating effect on the microwave emissions from the soil as was demonstrated by Schmugge *et al.* (1988). Aircraft experiments with an L-band push broom microwave radiometer over the Konza Prairie grasslands, Kansas, showed that for areas that had not been burned, the build-up of a thatch layer serves as a highly emissive layer above the soil, thus masking the emission of the soil itself. Where this thatch layer was absent because of burning or grazing, the microwave sensitivity to soil moisture was as expected for bare soil.

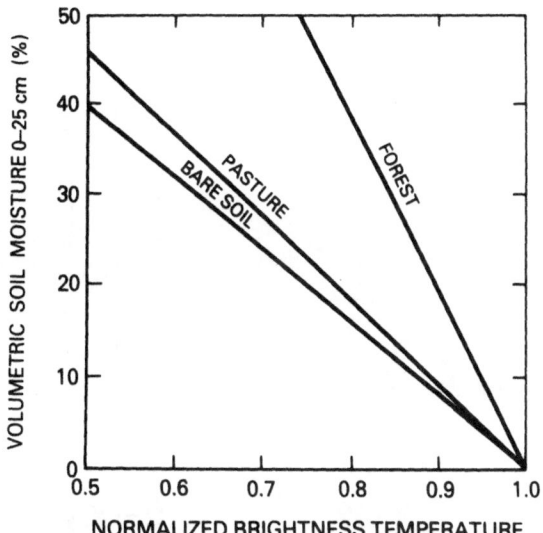

Figure 7.9 An illustration of the relationships between normalized brightness temperature and soil moisture for bare soil and different types of vegetation; L-band H at 10° (after Jackson *et al.* 1982).

Theis *et al.* (1984) demonstrated the use of visible and infrared data to calculate a perpendicular vegetation index (PVI), which in turn was used to correct the L-band emissivity determined with a passive microwave radiometer. They found that, as long as the PVI was less than 4.3, good results could be obtained. This research indicates the possibility of a total satellite remote sensing approach for soil moisture without any ground sampling.

Active microwave

Active microwave sensing of soil moisture faces the same potential and the same limitations as the passive case, but with some important differences. The effect of roughness and in some cases of the vegetation canopy can be more serious for radar. However, judicial choice of instrument parameters such as the angle of incidence, polarization and frequency can minimize these effects. These will be discussed in more detail below.

(a) Surface roughness. As with the passive case, surface roughness effects depend upon the wavelength, but they are also highly dependent upon the angle of incidence of viewing. As can be seen in Figure 7.10, the choice of an angle of incidence of less than 15° or 20° will minimize the effects of roughness. It can also be seen in Figure 7.10 that, for the roughest plot, the backscatter is almost independent of the angle of incidence.

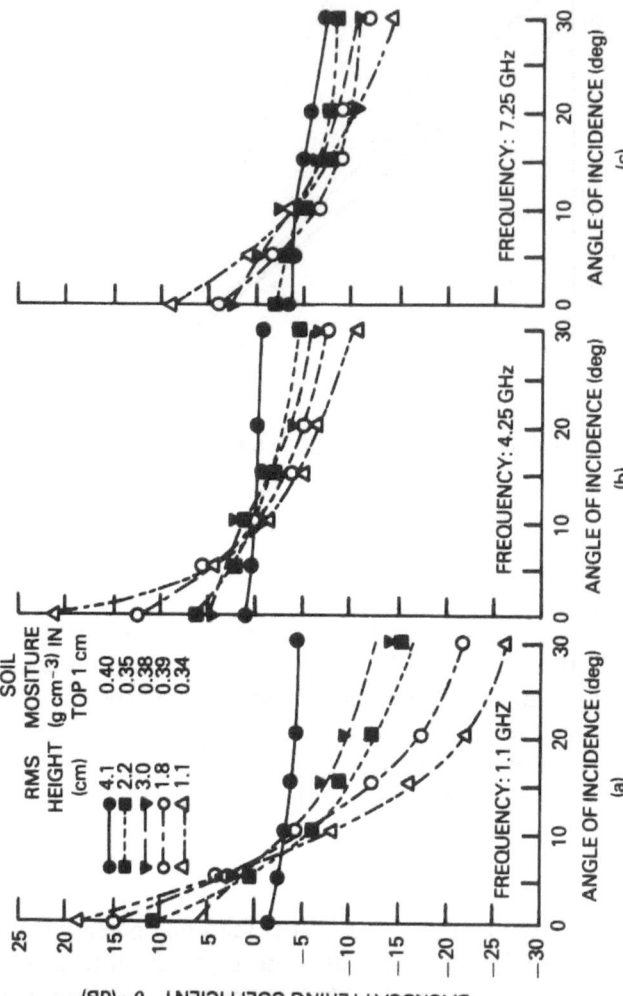

Figure 7.10 An illustration of the effect of angle of incidence on backscattering coefficient for different surface roughnesses: (a) 1.1 GHz, (b) 4.25 GHz, (c) 7.25 GHz (after Ulaby *et al.* 1986).

(*b*) *Vegetation effects.* As with the roughness case, the effect of vegetation on the active microwave sensing of soil moisture is greatly dependent upon the instrument angle of incidence, frequency and polarization. These effects are illustrated in Figure 7.11 for a corn canopy. In Figure 7.11(a) it can be seen that the attenuation for the horizontal polarization is relatively weak, but the vertically polarized data are attenuated to a much greater degree because of their relationship to the canopy structure, which consists primarily of vertical stalks. The effect of frequency on penetration depth can be seen in Figure 7.11(b). It is readily apparent that the penetration depth increases with a decrease in frequency or an increase in wavelength.

(*c*) Measurement depth. The same principles control the depth of soil that is being measured by the microwave technique, whether it is passive, as

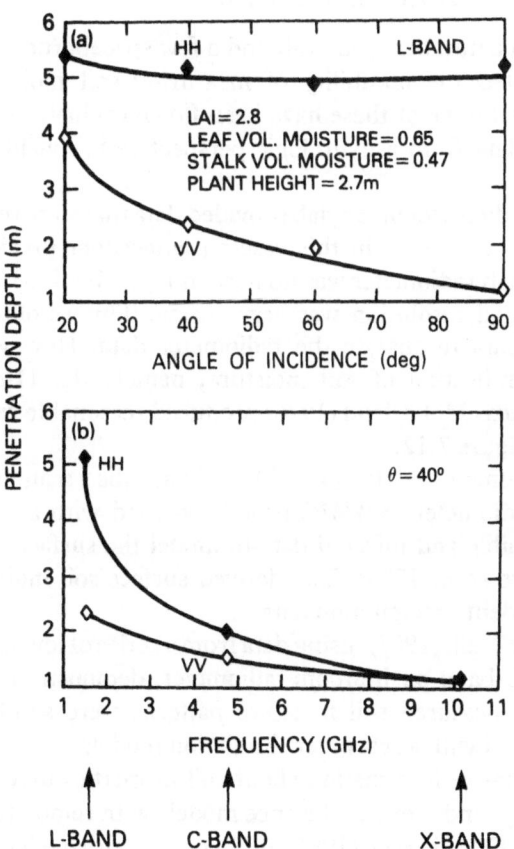

Figure 7.11 An illustration of the penetration depth of a corn canopy against angle of incidence for HH and VV polarization at (a) L-band and (b) 40° (after NASA 1988).

discussed above, or active. In a series of careful field experiments with a C-band, HH polarization radar, Bruckler *et al.* (1988) showed that the experimental results of penetration depth compared with soil moisture followed very closely the theoretical curve for a uniform profile.

Thus, the similarity between passive and active microwave techniques is evident. In the simplest terms, the passive systems appear to be less affected by surface roughness and overlying vegetation. Passive instrumentation would appear to be the obvious choice if it were not for a serious degradation in spatial resolution as the distance from the target increases. On the other hand, active systems using the synthetic aperture concept have a spatial resolution that is independent of sensor altitude. Consequently, the choice of technique becomes more difficult and based, at least on our current understanding, very much dependent upon the intended use of the data.

7.4 HYDROLOGIC APPLICATIONS

There have been a number of aircraft and a few spaceborne instruments that have demonstrated the feasibility of measuring soil moisture by remote sensing. Although none of these have been flown exclusively for hydrology, analysis of the data from a hydrologic perspective has yielded encouraging results.

The L-band radiometer on Skylab provided data that were related indirectly to a hydrologic parameter, in this case an antecedent precipitation index. Because the Skylab radiometer was non-scanning with a 110 km field of view, its resulting spatial resolution was very coarse. This resolution prohibited use of ground data to analyse the radiometer data. However, by using a more regional indication of soil moisture, namely the 11-day antecedent precipitation index, McFarland (1976) produced reasonable results which are reproduced in Figure 7.12.

Satellite radiometer data (18 GHz) from the scanning microwave multichannel radiometer (SMMR) have been used with a vegetation index derived from visible and infrared data to model the surface soil moisture in East Texas (Owe *et al.* 1988). The derived surface-soil moisture compared well with antecedent precipitation data.

Jackson and O'Neill (1987), using data from a series of low-elevation aircraft flights with an L-band push broom radiometer, demonstrated that observed and microwave measured soil moisture patterns were similar to temporal patterns developed with a soil water simulation model.

Useful quantitative information about soil properties may be obtained by calibrating energy and moisture balance models with remotely sensed passive microwave data. Camillo *et al.* (1986) used a soil physics model that is driven by the surface energy balance to solve the heat and moisture flux equations in the soil profile. Model-generated surface temperature and soil moisture and

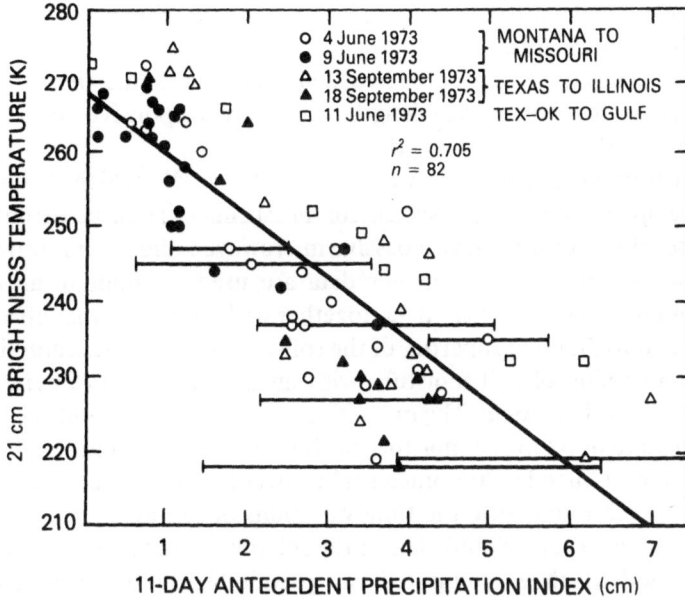

Figure 7.12 An illustration of L-band brightness temperatures obtained with a radiometer on board the Skylab spacecraft against the values of the antecedent precipitation index. The error bars represent the standard duration of the antecedent precipitation index for 6 to 10 stations within each resolution element (after Schmugge and McFarland 1976, 1983).

temperature profiles are then used in a microwave emission model to predict the soil brightness temperature. The model hydraulic parameters are then varied until the predicted temperatures agree with the remotely sensed measurements. Using this approach, values of hydraulic conductivity, matric potential, soil moisture at saturation and a soil texture parameter were estimated and found to agree well with measured and published values.

The utility of airborne gamma radiation flights for measuring soil moisture was demonstrated by an analysis of data from 19 flight lines over the Konza Prairie in 1987 (Carroll *et al.* 1988). The time series of the gamma-radiation-measured soil moisture to a depth of 20 cm was compared with an extensive set of ground data. The absolute error was within 2.3% of actual soil moisture.

Airborne radiometers have been used in the USSR for a number of hydrologic applications. Basharinov *et al.* (1979) have used passive microwave radiometers to estimate the presence of shallow groundwater. The problem of identifying areas of seepage from canals was shown to be possible with microwave measurements (Shutko 1981). Armand *et al.* (1979) and Basharinov *et al.* (1979) presented results of flight paths across rivers and flood plains with

30 and 5 cm wavelength radiometers. Their results demonstrated the potential capability of microwave remote sensing for identifying potentially active hydrologic zones (i.e. partial area hydrology) and for delineating floodplains. As expected, the 30 cm wavelength was more responsive to deeper soil moisture than the 5 cm wavelength.

Scientists in the USSR (Reutov and Shutko 1986; Mkvtchjan *et al.* 1988a,b) have developed an operational system for measuring soil moisture from aircraft which uses the output from two passive microwave radiometers (wavelengths of 18 and 30 cm). The radiometer data are used as input to an onboard microcomputer system. These data, together with prior information about the physical and hydraulic properties of the soils, allow them to accomplish near-real-time mapping of soil moisture with an accuracy of $0.05 \, \text{g cm}^{-3}$ if the biomass is less than about $2 \, \text{kg m}^{-3}$. The two wavelengths and the soil data enable them to develop soil moisture profiles to a depth of 1 m.

Data from a C-band scatterometer taken over a variety of agricultural fields have been used to measure the time variations of soil moisture over a 9-day period (Bernard *et al.* 1986b). Using the microwave data and a simple water budget model, they computed a pseudodiffusivity parameter that quantitatively describes the drying heterogeneities observed in the different fields.

Soares *et al.* (1987) investigated the spatial and temporal behaviour of C-band scatterometer and thermal infrared measurements over a wide variety of agricultural fields. They concluded that, at least for their study area, individual fields can be considered homogeneous with respect to surface temperature and surface soil moisture.

Jackson *et al.* (1986) demonstrated how soil moisture can be mapped with an airborne push broom radiometer. A $2700 \, \text{km}^2$ area of the Texas High Plains was mapped to assess the preplanting soil moisture. In a recent aircraft experiment in Kansas, a four-beam push broom microwave radiometer operating at $1.4 \, \text{GHz}$ (21 cm, L-band) was used to map the spatial distribution of soil moisture (Wang *et al.* 1989). By using overlapping flight lines for several flights during a drying period, the spatial patterns of soil moisture within a small 37.7 ha watershed were able to be mapped. The resulting maps are shown in Figure 7.13. The application of these maps and data to modelling runoff were discussed in Chapter 6.

Some early SAR results of soil moisture detection were presented by Ulaby *et al.* (1983) when they compared a SEASAT SAR image of an area near Waterloo, Iowa, with rain gauge data. This comparison, shown in Figure 7.14, shows a clear relationship between the bright return area in the eastern portion of the image and the rain gauge amounts resulting from a rain squall the previous day.

Soil moisture experiments with the Shuttle imaging radar-B were used to demonstrate the high-resolution capability of SAR from a space platform

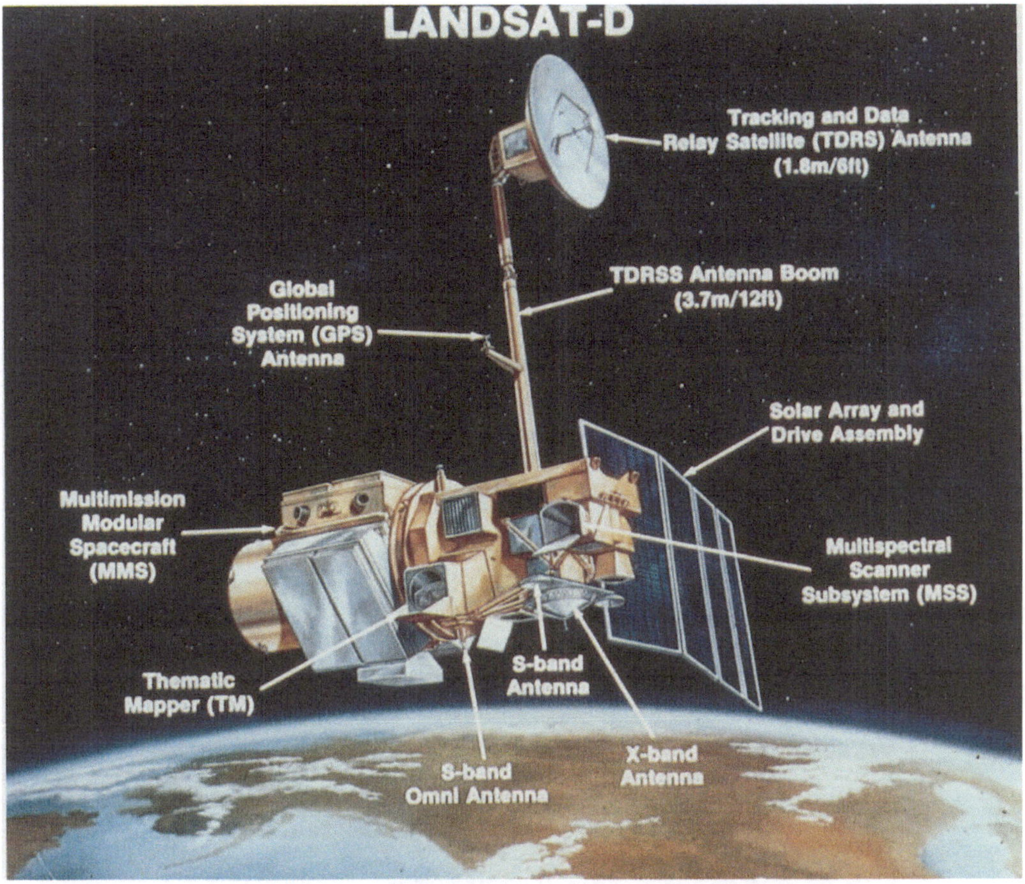

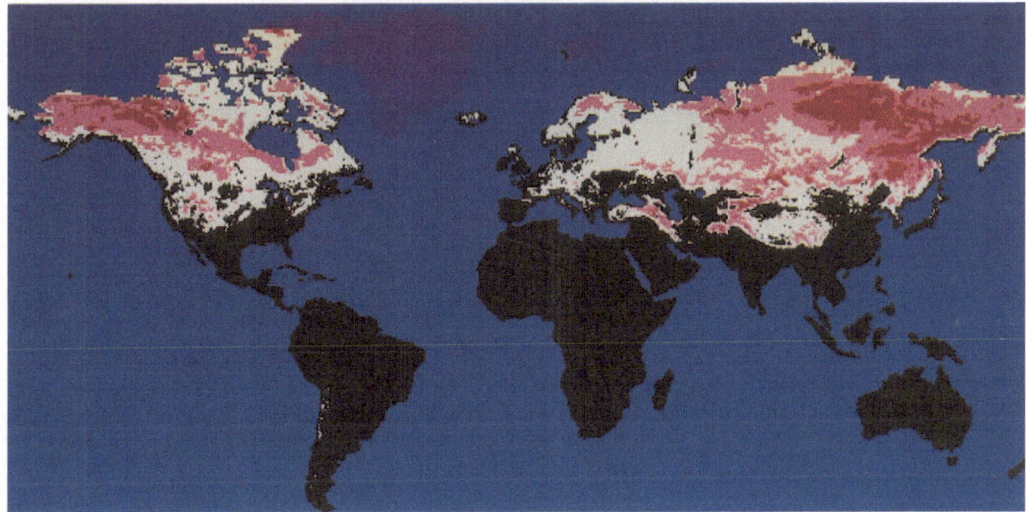

Plate 1 An artist's view of the Landsat 4 spacecraft showing the solar panel, instruments and communications antenna as deployed (D. Williams, GSFC).

Plate 2 An illustration of the snow-covered area of the World for February 1983. These data were obtained from the Nimbus-7 SMMR using data at 37 GHz and are colour coded to show snow depth (courtesy of Chang *et al.* 1986).

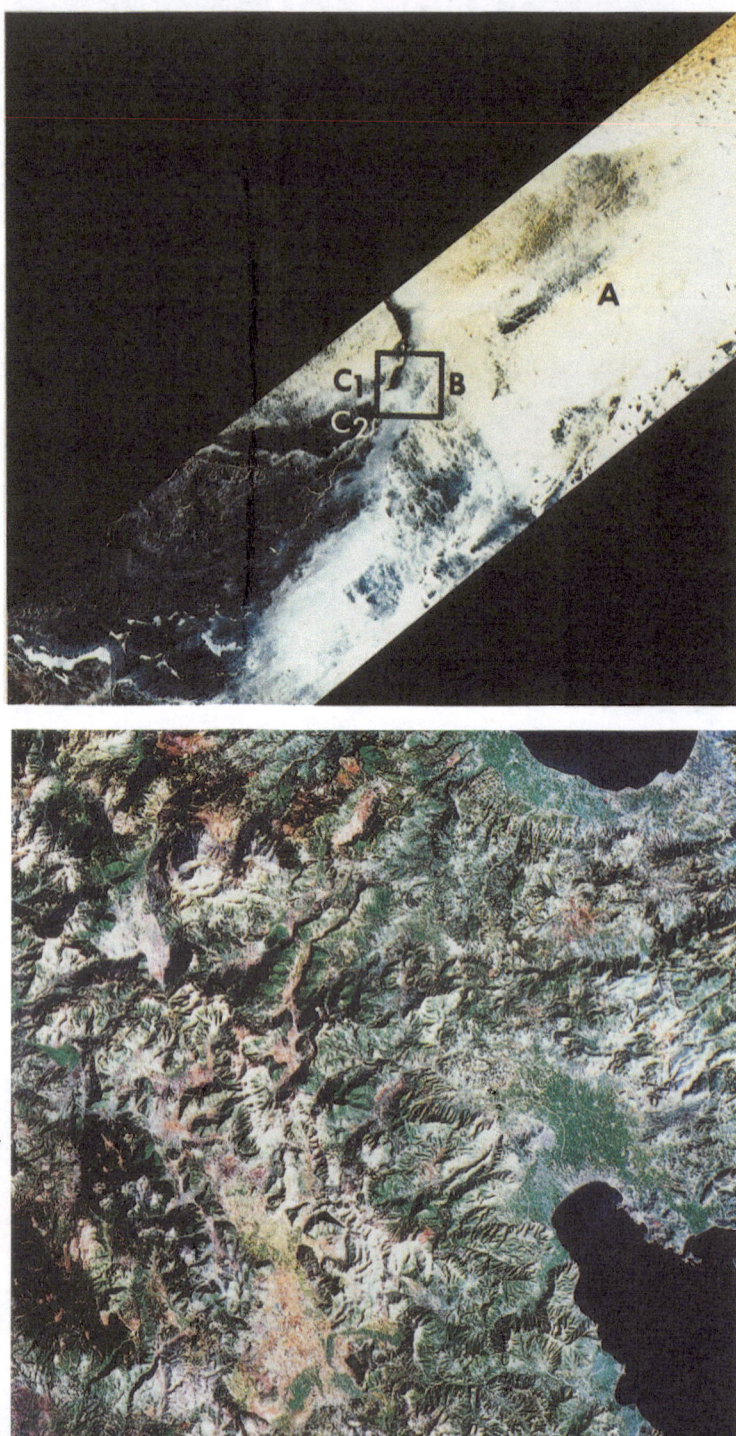

Plate 4 Colour composite of the Gulf of Argos area (bands 1, 4, 7), size 2500 × 2500 pixels, used by Kaufman *et al.* (1986) to identify different lithological features on the basis of texture and vegetation.

Plate 3 (opposite) Coregistered Landsat MSS and Seasat data for the Al Hisma region of Saudi Arabia, showing how the data can be used for geological interpretation (courtesy Chavez *et al.* 1983). A, drift sand deposits; B, wadi deposits; C, Tawil sandstone and Tabuk formation.

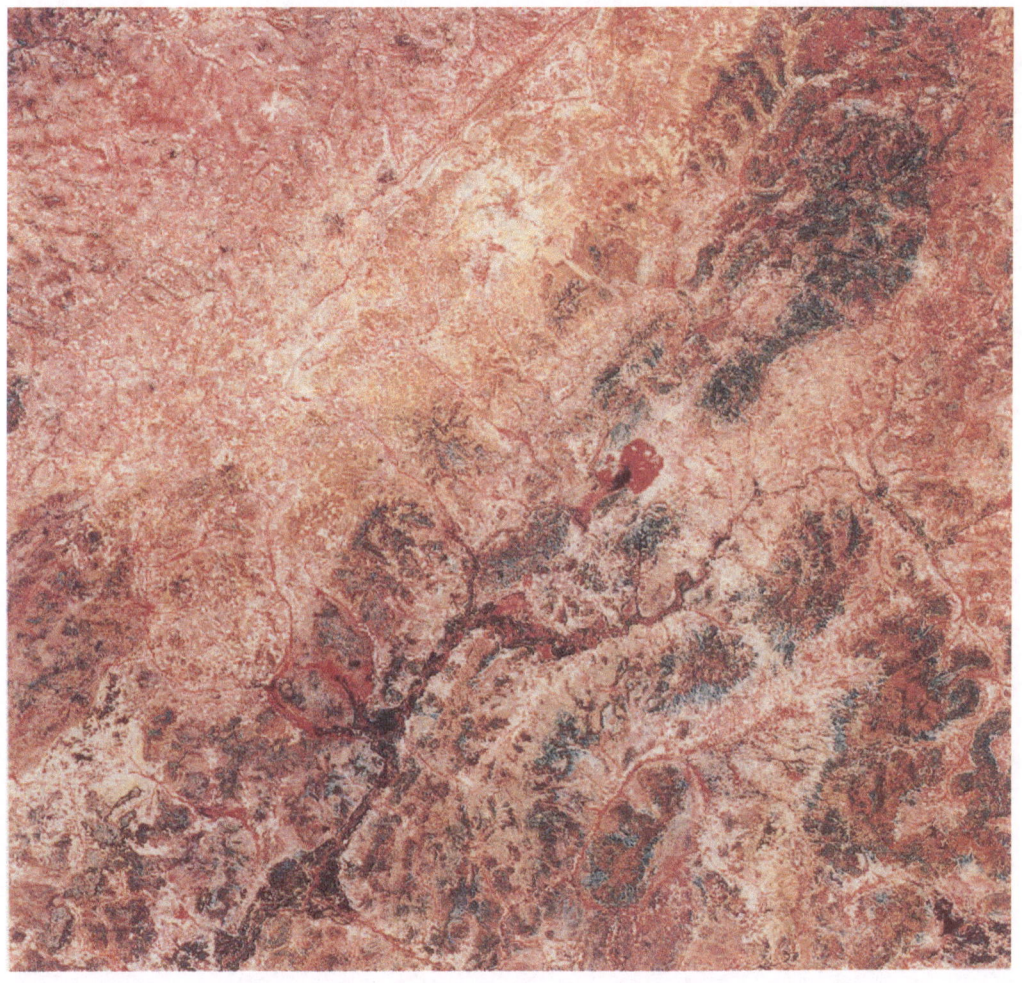

Plate 5 Computer-enhanced colour composite, 1 500 000 scale Landsat image of the Bobo-Dieulasso area, Upper Volta, West Africa. Acquired 31 March 1976 and processed by Earth Satellite Corporation to show geological structure (see also Figure 8.6; reproduced with permission from the American Water Resources Association.)

Plate 6 Composite of Figures 8.8(a) and (b) showing enhancement of surface detail and structure using synergism of Seasat SAR and Landsat MSS (Chavez and Sanchez 1981).

Plate 7 A false-colour image of thematic mapper bands 1, 2 and 3 for Lake Chicot in Arkansas. The decrease in suspended sediment along the lake is evident by the decrease in reflectance (bright colour) and increase in the dark colour signifying clear water (courtesy of Jerry C. Ritchie, USDA).

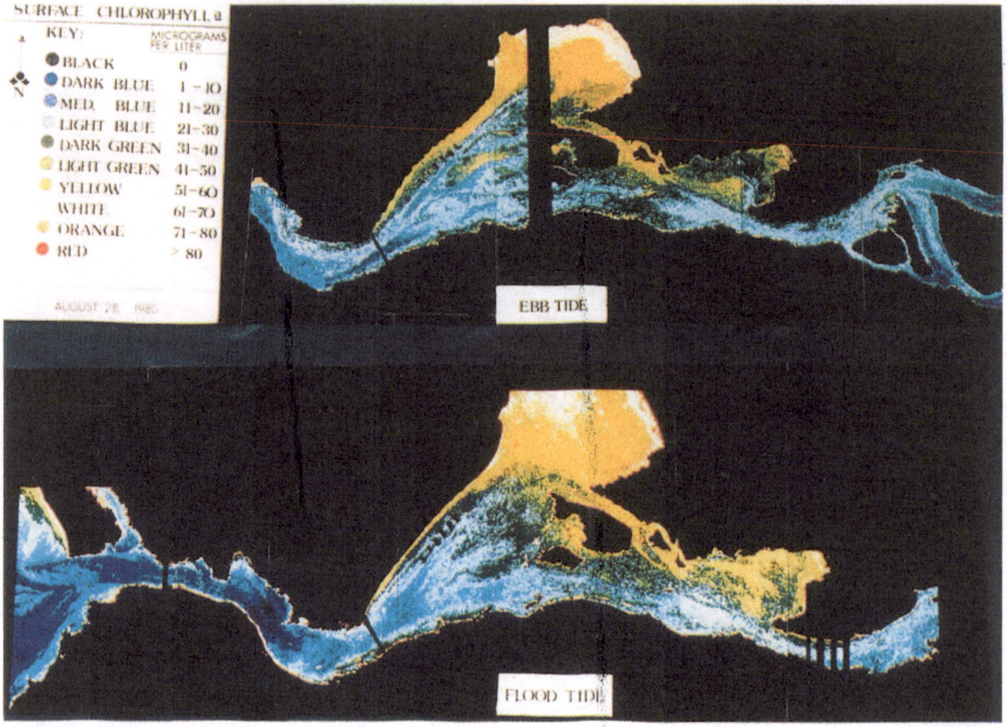

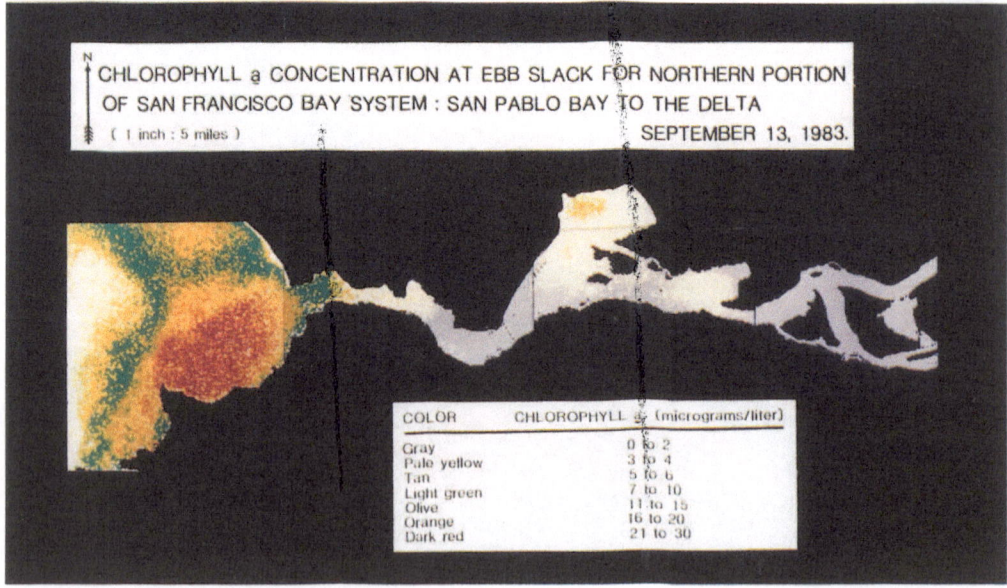

Plate 8 Predicted spatial patterns of surface chlorophyll *a* concentration based on surface data acquired within 15 mm of Daedalus 1260 MSS overflight: (a) morning and afternoon data of 28 August 1980; (b) morning data of 13 September 1983 (courtesy of Khorram *et al.* 1987; reproduced with permission from IEEE).

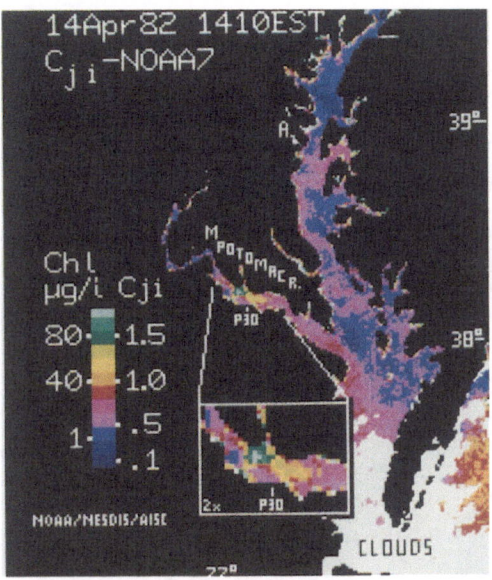

Plate 9 Relative pigment concentrations for the Chesapeake Bay using a colour index derived from from NOAA 7 AVHRR, April 1982. Inset shows bloom in lower Potomac River (courtesy of Stumpf and Tyler 1988).

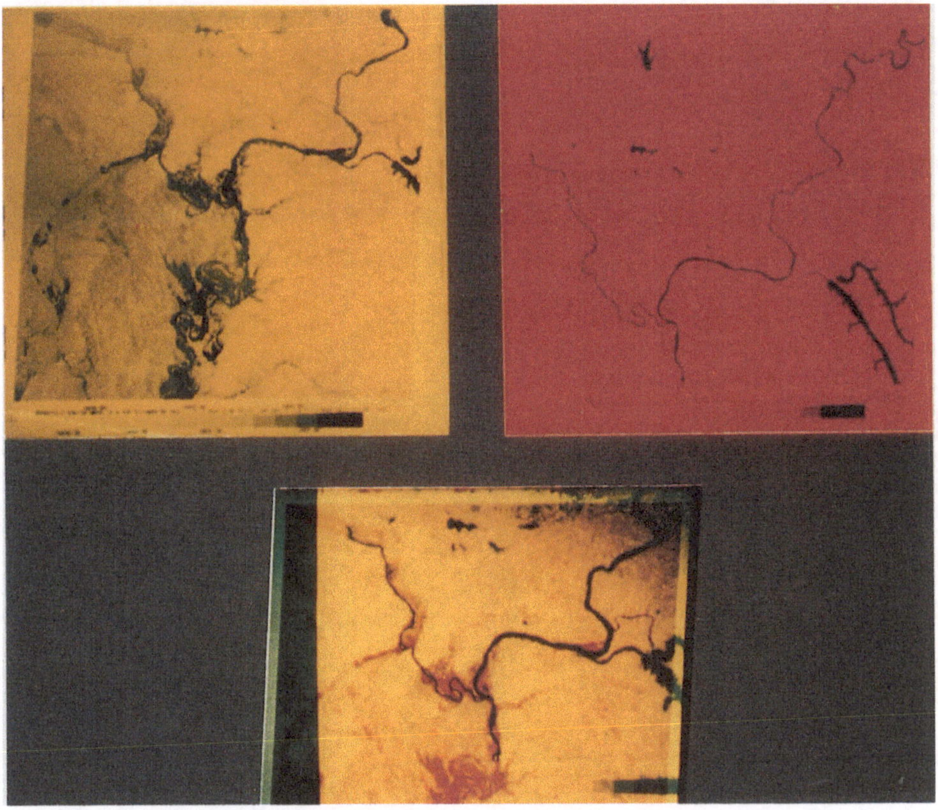

Plate 10 An illustration of a composite flood scene superimposed over a pre-flood scene, Landsat band 7 (courtesy of Deutsch and Ruggles 1974).

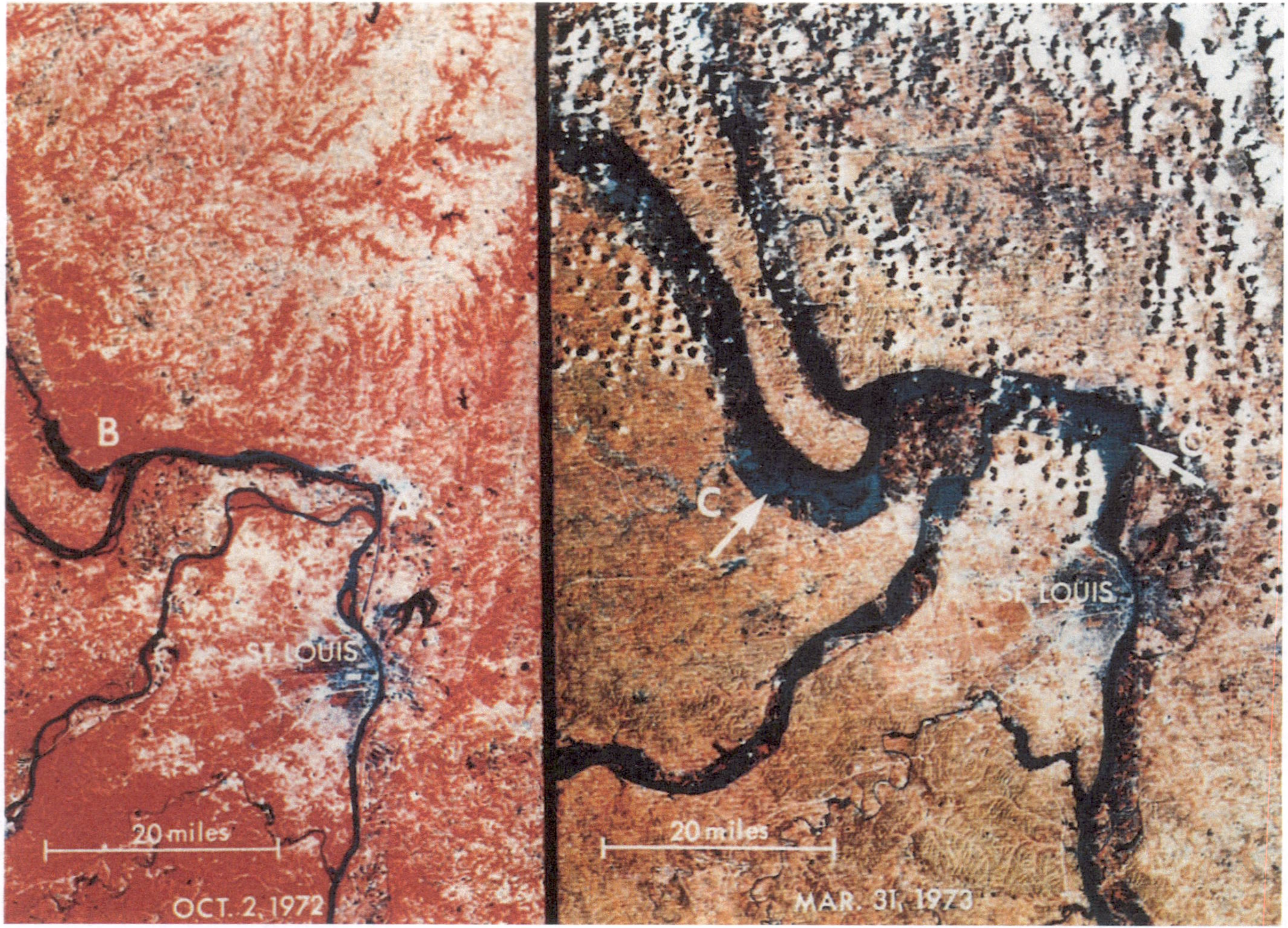

Plate 11 A comparison of pre- and post-flooding of the St Louis, Missouri, area for the 1973 flooding of the Mississippi River (courtesy of Rango and Salomonson 1974).

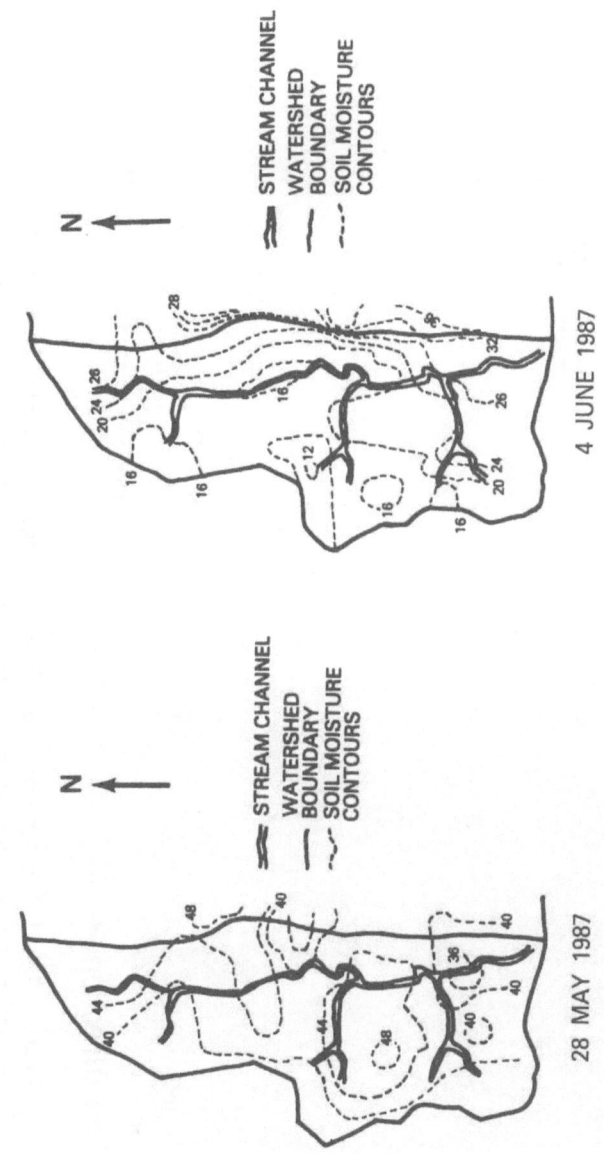

Figure 7.13 An illustration of the temporal and spatial changes in soil moisture mapped by a push broom microwave radiometer over a 37.7 ha watershed (after Wang et al. 1988).

f = 1.275 GHz
θ = 23 ± 3 Degrees
Polarization: HH

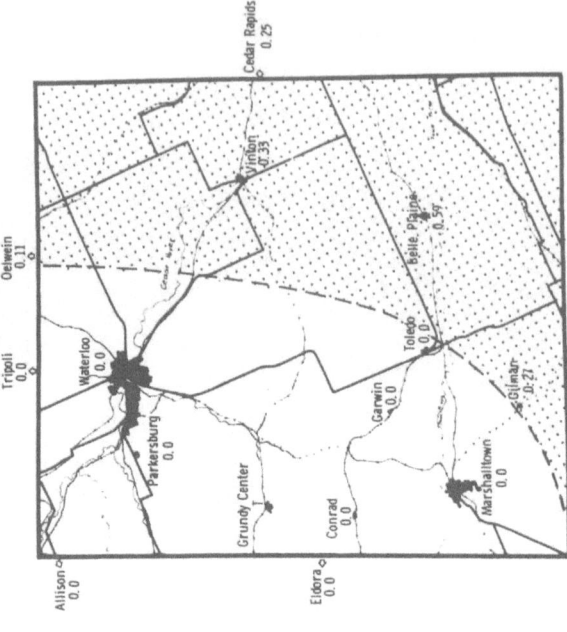

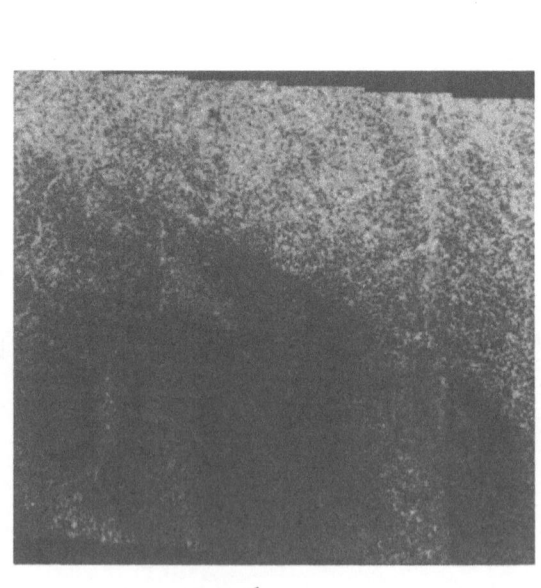

Illumination Direction ⟶

Figure 7.14 A comparison of a SEASAT image and ground-based rainfall observations for the area around Waterloo, Iowa (rainfall in inches) (courtesy of Ulaby *et al.* 1983; reproduced with permission from IC22).

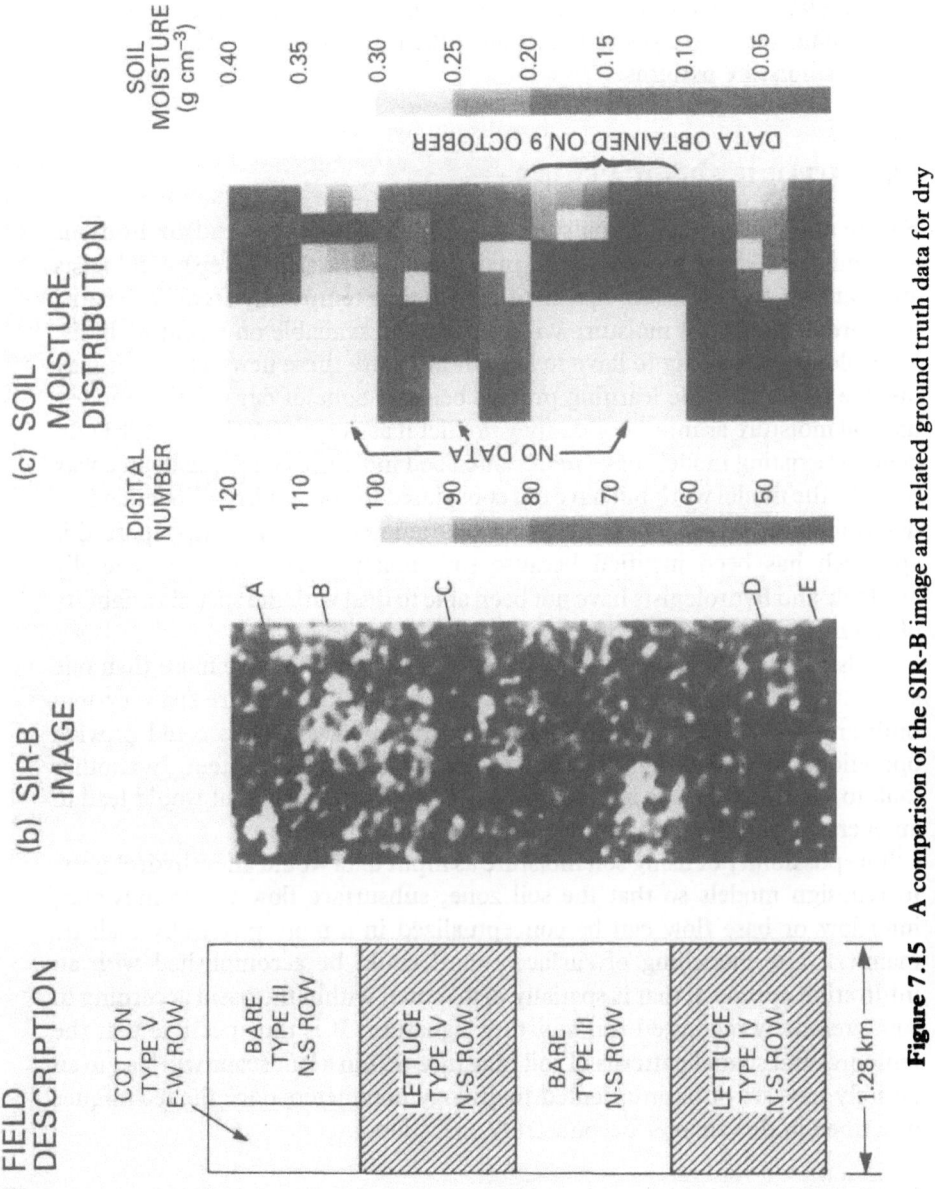

Figure 7.15 A comparison of the SIR-B image and related ground truth data for dry and irrigated fields (after Wang *et al.* 1986).

(Wang *et al.* 1986). Figure 7.15 shows the SAR image and the corresponding ground data for some irrigated lettuce fields. Portions of the fields were undergoing various stages of irrigation at the time, which explains the highly variable moisture patterns.

7.5 FUTURE APPLICATIONS

Future applications of soil moisture to hydrologic questions and applications are bound to become more common in the coming years. One step hydrologists are going to have to overcome is to realize that temporally frequent spatial measurements of soil moisture will someday be available on a routine basis. Hydrologists are going to have to learn how to use these new data. This may involve a considerable learning process because none of our existing models use soil moisture as input nor do they predict it as output (Peck *et al.* 1981). In general, existing models have represented soil moisture conceptually in a way to make the model work but have not considered the possibility of independent determination of soil moisture or soil parameters. For the most part this approach has been justified because soil moisture data are not generally available and hydrologists have not been able to deal with the spatial variability of soil moisture and soil properties.

It also appears that operational remote sensing will involve more than one sensor. Bernard *et al.* (1986a) have demonstrated how a future space system built around visible, thermal infrared and microwave sensors could provide operational information of the soil/vegetation/atmosphere system. In another look to the future, Jackson (1988) reviewed previous work that would lead to an operational passive microwave system for soil moisture.

The possibility of using soil moisture as input data would allow hydrologists to redesign models so that the soil zone, subsurface flow to groundwater, interflow or base flow can be conceptualized in a more physically realistic manner. The modelling of surface runoff could be accomplished with an infiltration approach that is spatially distributed within the basin according to some remotely measured soil moisture signature. It is thus possible that the temporal and spatial patterns of soil moisture within a landscape will lead to an entirely new set of basin-oriented hydrology parameters once the techniques described in this chapter become truly operational.

REFERENCES

Armand, N. W., Basharinov, A. E. and Shutko, A. M. (1979) Recent microwave radiation studies of continental covers. *Acta Astronaut.* **6**, 647–55.

Barton, I. J. (1978) A case study of microwave radiometer measurements over bare and vegetated surfaces. *J. Geophys. Res.* **83**, 3513–17.

Basharinov, A. E., Krylova, M. S., Maslov, A. I., Shutko, A. M. (1979) Remote sensing of subsurface soil moisture by means of microwave radiometers. *Water Resour.* 5, 538–42.

Basharinov, A. Y. and Shutko, A. M. (1975) Simulation studies of the SHF radiation characteristics of soils under moist conditions, *NASA Tech. Trans.* TTF-16.

Batlivala, P. P. and Ulaby, F. T. (1977) Estimation of soil moisture with remote sensing. *ERIM Proc. 11th Int. Symp. on Remote Sensing of Environment,* vol. 2, pp. 1557–66.

Bernard, R., Taconet, O. and Vidal-Madjar, D. (1986a) Toward a satellite system to monitor the spatial and temporal behavior of the soil water content. *Proc. IGARSS'86 Symp, Zürich,* pp. 751–3.

Bernard, R., Soares, J. V. and Vidal-Madjar, D. (1986b) Differential bare field drainage patterns from airborne microwave observations. *Water Resour. Res.* 22, 869–75.

Bruckler, L., Witono, H. and Stengel, P. (1988) Near surface soil moisture estimation from microwave measurements. *Remote Sensing Environ.* 26, 101–21.

Camillo, P. J., O'Neill, P. E. and Gurney, R. J. (1986) Estimating soil hydraulic parameters using passive microwave data. *IEEE Trans. Geosci. Remote Sensing,* GE-24, 930–6.

Carroll, T. R. (1981) Airborne soil moisture measurements using natural terrestrial gamma radiation. *Soil Sci.* 132, 358–66.

Carroll, T. R., Peck, E. L. and Lipinski, D. M. (1988) Airborne time-series measurement of soil moisture using terrestrial gamma radiation. *Proc. Annu. Convention, Am. Congr. on Surveying and Mapping and Am. Soc. Photogram. and Remote Sensing,* St Louis, MO.

Choudhury, B. J., Schmugge, T. J., Newton, R. W. and Chang, A. (1979) Effect of surface roughness on the microwave emission from soils. *J. Geophys. Res.* 81, 3660–6.

Crist, E. P. and Cicone, R. C. (1984) Physically based transformation of thematic mapper data: VTM tasseled cap. *IEEE Trans. Geosci. Remote Sensing.* GE-22, 256–63.

Engman, E. T. (1982) Remote sensing applications in watershed modeling. In *Applied Modeling in Catchment Hydrology,* Water Resources Publications, Littleton, CO, pp. 473–94.

Idso, S. B., Schmugge, T. J., Jackson, R. D. and Reginato, R. J. (1975) The utility of surface temperature measurements for remote sensing of surface soil water status. *J. Geophys. Res.* 80, 3044–9,

Jackson, R. D., Ahler, J., Estes, J. E., Heilman, J. L., Kakle, A., Kanemasu, E. T., Millard, J., Price, J. C. and Wiegand, C. (1978) Soil moisture estimation using reflected solar and emitted thermal radiation. *Soil Moisture Workshop, Chapter 7,* NASA CP-2073.

Jackson, T. J. (1988) Research toward an operational passive microwave remote sensing system for soil moisture. *J. Hydrol.* 102, 95–112.

Jackson, T. J., Hawley, M. E., Shiue, J., O'Neill, P. E., Owe, M., Delnore, V. and Lawrence, R. W. (1986) Assessment of preplanting soil moisture using airborne microwave sensors. *Hydrologic Applications of Space Tech., IAHS Publ. No. 160,* pp. 111–18.

Jackson, T. J. and O'Neill, P. (1987) Temporal observations of surface soil moisture using a passive microwave sensor. *Remote Sensing Environ.* **21**, 281–96.

Jackson, T. J. and Schmugge, T. J. (1986) Passive microwave remote sensing of soil moisture. *Adv. Hydrosci.* **14**, 123–59.

Jackson, T. J. and Schmugge, T. J. (1989) Passive microwave remote sensing system for soil moisture: Some supporting research. *IEEE Trans. Geosci. Remote Sensing* **GE-27**, 225–35.

Jackson, T. J., Schmugge, T. J. and O'Neill, P. (1984) Passive microwave remote sensing of soil moisture from an aircraft platform. *Remote Sensing Environ.* **14**, 135–51.

Jackson, T. J., Schmugge, T. J. and Wang, J. R. (1982) Passive microwave sensing of soil moisture under vegetation canopies. *Water Resour. Res.* **18**, 1137–42.

Kirdiashev, K. P., Chukhlantsev, A. A. and Shutko, A. M. (1979) Microwave radiation of the Earth's surface in the presence of vegetative cover. *Radio Eng. Electron.* (English transl.), **24**, 256–64.

Mkvtchjan, F. A., Reutov, E. A., Shutko, A. M., Kostov, K. G., Michalev, M. A., Nedeltchev, N. M., Spasov, A. Y. and Vichev, B. I. (1988a) Microcomputer-based radiometer data acquisition and processing system for large-area mapping of soil moisture content in the top one meter layer. *Proc. IGARSS'88 Symp., Edinburgh, Scotland*, ESA SP-284, pp. 1563–4.

Mkrtchjan, F. A., Reutov, E. A., Shutko, A. M., Kostov, K. G., Michalev, M. A., Nedeltchev, N. M., Spasov, A. Y. and Vichev, B. I. (1988b) Experiments in Bulgaria for determination of soil moisture in the top one meter layer using microwave radiometry and a priori information. *Proc. IGARSS'88 Symp., Edinburgh, Scotland*, ESA SP-284, pp. 665–6.

McFarland, M. J. (1976) The correlation of SKYLAB L-band brightness temperatures with antecedent precipitation. *Conf. on Hydrometeorology, Preprints, Fort Worth, TX*, pp. 60–5.

Mo, T., Schmugge, T. J. and Choudhury, B. J. (1980) Calculations of the spectral nature of the microwave emission from soils. *NASA Tech. Memo.* 82002.

NASA (1988) *SAR Instrument Panel Report, Earth Observing System*, IIf.

Newton, R. W., Black, Q. R., Makanvand, S., Blanchard, A. J. and Jean, B. R. (1982) Soil moisture information and thermal microwave emission. *IEEE Trans. Geosci. Remote Sensing* **GE-21**, 300–7.

Newton, R. W. and Rouse, J. W. (1980) Microwave radiometer measurements of moisture content. *IEEE Trans. Antenna Propag.* **AP-28**, 680–6.

Owe, M., Chang, A. and Golus, R. E. (1988) Estimating soil moisture from satellite microwave measurements and satellite derived vegetation index. *Remote Sensing Environ.* **24**, 331–45.

Peck, E. L., Keefer, T. N. and Johnson, E. R. (1981) Strategies for using remotely sensed data in hydrologic models. *NASA Rep. No.* CR-66729, Goddard Space Flight Center, Greenbelt, MD.

Perry, E. M. and Carlson, T. N. (1988) Comparison of active microwave soil water content with infrared surface temperatures and surface moisture availability. *Water Resour. Res.* **24**, 1818–24.

Price, J. C. (1982) Estimation of regional scale evapotranspiration through analysis of satellite thermal-infrared data. *IEEE Trans. Geosci. Remote Sensing* **GE-20**, 286–92.

Promes, P. M., Jackson, T. J. and O'Neill, P. E. (1988) Significance of agricultural row structure on the microwave emissivity of soils. *IEEE Trans. Geosci. Remote Sensing* **26**, 580–9.

Reutov, E. A. and Shutko, A. M. (1986) Prior-knowledge-based soil-moisture determination by microwave radiometry. *Sov. J. Remote Sensing* **5**, 100–25.

Schmugge, T. J. (1983) Remote sensing of soil moisture: Recent advances. *IEEE Trans. Geosci. Remote Sensing* **GE-21**, 336–344.

Schmugge, T. J., Jackson, T. J. and McKim, H. L. (1980) Survey of methods for soil moisture determination. *Water Resour. Res.* **16**, 961–79.

Schmugge, T. J., Njoku, E. G., Peck, E. and Ulaby, F. T. (1978) Microwave and gamma radiation observations of soil moisture. *Soil Moisture Workshop*, Chapter 5, NASA, CP-2073.

Schmugge, T. J., Wang, J. R. and Asrar, G. (1988) Results from the push broom microwave radiometer flights over the Konza Prairie in 1985. *IEEE Trans. Geosci. Remote Sensing* **GE-26**, 590–6.

Shutko, A. M. (1981) Microwave radiometry of lands under natural and artificial moistening. *Paper presented at Symp. on Signature Problems in Microwave Remote Sensing, Union Radio Sci. Int., University of Kansas, Lawrence, KS.*

Soares, J. V., Bernard, R. and Vidal-Madjar, D. (1987) Spatial and temporal behavior of a large agricultural area as observed from airborne C-band scatterometer and thermal infrared radiometer. *Int. J. Remote Sensing*, **8**, 981–96.

Theis, S. W., Blanchard, B. J. and Blanchard, A. J. (1986) Utilization of active microwave roughness measurements to improve passive microwave soil moisture estimates over bare soils. *IEEE Trans. Geosci. Remote Sensing*, **GE-24**, 334–9.

Theis, S. W., Blanchard, B. J. and Newton, R. W. (1984) Utilization of vegetation indices to improve microwave soil moisture estimates over agricultural lands. *IEEE Trans. Geosci. Remote Sensing* **GE-22**, 490–6.

Ulaby, F. T., Brisco, B. and Dobson, M. C. (1983) Improved spatial mapping of rainfall events with spaceborne SAR imagery. *IEEE Trans. Geosci. Remote Sensing*, **GE-21**, 118–21.

Ulaby, F. T., Moore, R. K. and Fung, A. K. (1982) *Microwave Remote Sensing: Active and Passive*, Addison-Wesley, Reading, MA.

Ulaby, F. T., Moore, R. K. and Fung, A. K. (1986) *Microwave Remote Sensing: Active and Passive*, vol. III, Artec House, Dedham, MA.

van de Griend, A., Camillo, P. J. and Gurney, R. J. (1985) Discrimination of soil physical parameters, thermal inertia and soil moisture from diurnal surface temperature fluctuations. *Water Resour. Res.* **21**, 997–1009.

Wang, J. R., Engman, E. T., Shiue, J. C., Ruzek, M. and Steinmeier, C. (1986) The SIR-B observations of microwave backscatter dependence on soil moisture, surface roughness and vegetation covers. *IEEE Trans. Geosci. Remote Sensing*, **GE-24**, 510–16.

Wang, J. R., Newton, R. W. and Rouse, J. W. (1980) Passive microwave remote sensing of soil moisture: the effect of tilled row structure. *IEEE Trans. Geosci. Remote Sensing*, **GE-18**, 296–302.

Wang, J. R. and Schmugge, T. J. (1980) An empirical model for the complex dielectric permittivity of soils as a function of water content. *IEEE Trans. Geosci. Remote Sensing*, **GE-18**, 288–95.

Wang, J. R., Shiue, J. C., Schmugge, T. J. and Engman, E. T. (1989) Mapping soil moisture with L-band radiometric measurements. *Remote Sensing Environ.*, **27**, 305–12.

Wetzel, P. J., Atlas, D. and Woodward, R. H. (1984) Determining soil moisture from geosynchronous satellite infrared data: A feasibility study. *J. Clim. Appl. Meteorol.*, **23**, 375–91.

Wilheit, T. T. (1978) Radiative transfer in a plane stratified dielectric. *IEEE Trans. Geosci. Remote Sensing*, **GE-16**, 138–43.

8

Groundwater

8.1 INTRODUCTION

Groundwater refers to all water stored beneath the surface of the Earth in aquifers. In order to use the water stored in an aquifer, it is necessary first to locate the aquifer, map its size, extent and depth, and then estimate recharge and discharge rates of water. The effects of additional extraction wells on the aquifer and its structural integrity must also be assessed. Much of this information can only be confirmed by the use of well records, bore hole and other *in situ* sampling methods. However, the use of remote sensing techniques is a very cost-effective approach in prospecting, and in preliminary survey, because the cost of drilling is such that it is not cost effective to drill randomly. Remote sensing techniques in groundwater exploration are addressed below.

The wavelengths of electromagnetic radiation commonly used in remote sensing are either emitted or reflected from the surface or a relatively shallow layer (less than 1 m) of the Earth. Because of this, deep groundwater aquifers cannot be detected directly. However, remote sensing techniques offer the hydrogeologist a powerful tool to add to his standard geophysical methods. The interpretation of aerial photographs and satellite imagery enables geologists to infer the location of aquifers from surface features. Satellite imagery, in particular, enables image analysts to view very large areas and achieve a perspective not possible from ground surveys or even low-level aerial photography. Radar can be used to show surface features even under dense vegetation canopies and is especially valuable for revealing topographic relief and roughness.

8.2 GENERAL APPROACH

Aerial photography has been used for many years for preliminary surveys (Ray 1960; Nefodov and Popova 1972) and has recently been supplemented by satellite data, particularly from the Landsat series (Gurney *et al.* 1982). Satellite imagery provides a synoptic view of a region that allows large-scale features to be identified that otherwise may be missed with aerial photography or ground surveys (Moore 1979). These data permit inferences to be made about rock types, structure and stratigraphy. When combined with field surveys, the satellite data permit an interpolation between ground sampling points that helps define the surface structure better and more economically than by surface methods alone. In general, the analysis of aerial photography or satellite imagery should precede ground surveys and fieldwork, because it may eliminate areas of potentially low-water-bearing strata and may also indicate promising areas for intensive fieldwork (Revzon *et al.* 1983). How much information can be obtained from these types of remotely sensed data varies considerably, depending on the geology, climate and type of cover. In semi-arid and arid regions the geological features can usually be interpreted easily because there is little surface cover to hide the structural and stratigraphic information. Even where there is vegetative cover there is usually a close association between that cover and the underlying geology. Much of the interpretation of aerial photography and satellite data is subjective and dependent on ground and subsurface confirmation.

Structural features such as faults, fracture traces and other linear features can indicate the possible presence of groundwater. Similarly other features, such as sedimentary strata or certain rock outcrops, may indicate potential aquifers. Since the remote sensing approach is limited to surface features, the first step in groundwater exploration is the delineation of surface features and land forms.

Shallow groundwater can be inferred by soil moisture measurements and by changes in vegetation types and patterns. Groundwater recharge and discharge areas within drainage basins can be inferred from soils, vegetation and shallow or perched groundwater.

Differences in temperature measured by remote sensing have been used to infer or identify shallow groundwater and springs or seeps. These temperature differences are the result of the high heat capacity of the groundwater that produces a heat sink in the summer and a heat source in the winter. The near-surface soil, soil moisture and vegetation temperatures respond rapidly to the local meteorology, whereas the groundwater temperature changes are both diurnally and seasonally damped.

Synthetic aperture radar (SAR) data have a great potential for groundwater exploration, especially in arid and hyperarid regions. The penetrating capability of long-wave radar, and the capability of radar to detect soil moisture, make SAR a very valuable tool for exploration in arid regions. Side-looking radar (SLAR) mounted on aircraft or satellites has been successfully used to map and identify structural features in regions of the world that because of either perpetual cloud cover or thick vegetation have never been adequately mapped. They have also been used with Landsat to help interpretation (Plate 3).

8.3 EXPLORATION WITH SATELLITE IMAGERY

Satellite images contain geological and hydrologic information that must be inferred by analysis and interpretation. Groundwater information can be inferred from land-forms, drainage patterns, vegetation characteristics, land use patterns, linear and curvilinear features, and image tones and textures. In arid regions, vegetation characteristics may indicate groundwater depth and quality. Satellite data may be the only source of data in regions of the world where maps, geological surveys and other information do not exist or are not accurate.

Satellite imagery can be used most effectively for regional exploration. Analysis of satellite imagery is a rapid and inexpensive means of obtaining reconnaissance groundwater information. The large areal coverage and relatively coarse spatial resolution are more suitable for obtaining general aquifer information than for siting test wells. The results of Landsat or SPOT image interpretation can result in:

1. less need for fieldwork and slower, more expensive exploration methods;
2. identification of promising areas for more detailed study and ground exploration;
3. new or better geological and hydrologic information;
4. the perspective of large areal coverage available from satellite imagery, which may be unavailable from other means of exploration.

It is important to emphasize, however, that remote sensing is nothing more than an additional source of information. It does not replace the more traditional techniques used by groundwater hydrologists such as topographic maps, seismic and resistance surveys, ground-penetrating radar, etc. — it merely supplements them. Nevertheless, there are some unique features of satellite imagery that make it extremely valuable.

8.4 PRINCIPLES OF IMAGE ANALYSIS

Satellite images are essentially pictures of the land surface. An image is composed of elements that represent the physical, structural biological and cultural features of the landscape. These are the clues used in image analysis.

Groups of similar elements may represent similar hydrogeological conditions. It is the job of the groundwater hydrologist to detect, delineate, classify and identify all groups that may have geological or hydrologic significance. For this purpose the groundwater hydrologist relies on the well-proven principles of photograph interpretation and utilizes characteristics such as tone or colour, texture, pattern, size, shape, location, elevation and association.

The result of image analysis is an interpretation in which landscape characteristics have been classified into several categories.

1. Land forms are considered recognizable physical features on the Earth's surface (i.e. bedrock mountains, volcanic features, alluvial fans, glacial features). For example, groundwater can be assumed to move down the lower slopes and down alluvial fans in the same direction as the surface streams. In large basins, it would be assumed that coarse-grained materials that were transformed by older drainage systems are located farther out and covered by finer-grained and less permeable materials. More recent coarse-grained sediments would be at the edges of the basin.
2. Drainage features include basin size and shape, drainage patterns and density, valley shape, channel location and angles of tributaries. The assumption is that joints and faults in the bedrock influenced the development of drainage patterns.
3. Cover is defined as various types of vegetation (both natural and man induced) and soils. Dense vegetation in valleys or basins indicates the availability of adequate water where the groundwater may be close to the surface. Generally, riparian vegetation and phreatophytes show up as bright red on false-colour imagery whereas xerophytic plants are more likely to appear brownish and sparsely distributed.
4. Lineaments are straight to slightly curving lines form in many different types of landscape. Many linear features are not continuous but need to be extended or joined in image analysis. It is assumed that lineaments mark the location of joints and faults.
5. Curvilinears are symmetrical lines with circular, elliptical or accurate shapes. Similar to lineaments in analysis.
6. Texture is manifested in the density of the drainage patterns. Fine-texture drainage systems indicate fine-grained sediments and areas where infiltration is relatively slow. Maximum infiltration would occur along major streams and in areas characterized by medium- and coarse-textured drainage.

8.5 IMAGERY SELECTION

The criteria used in image selection for groundwater investigations may differ from remote sensing applications. Proper selection of imagery can greatly facilitate groundwater exploration. In the case of Landsat, TM or SPOT imagery, the proper choice of spectral bands and time of year can be important considerations. The following rules of thumb should aid in the selection of imagery.

1. Select scenes with low sun angle. Land forms and general topography are enhanced by topographic shadowing when the sun angle is relatively low (less than 45°).
2. Select black-and-white infrared images. With Landsat, the band 7 images are good for delineating landscape features without being confused by vegetation tones.
3. Select at least one false-colour scene. False-colour composites show landforms and drainage patterns with respect to vegetation types and patterns, and maximize the differences between types of vegetation. The vegetation patterns and brightness are clues to the location and proximity of groundwater. Dry-season scenes will show phreatophytes as bright red whereas plants without adequate moisture will appear dull red or brown.
4. Select two black-and-white images from different orbits. Using images from different orbits (and dates) can provide a rough stereoscopic view of the study area. Orbital drift or cutting the scenes from slightly different locations along the orbital track provides this stereoscopic effect. It is not controlled or precise enough to map elevation differences but it can be useful for identifying slopes and land forms. This is particularly useful with SPOT data, where the sensor data can be obtained other than in the vertical.

8.6   CURRENT APPLICATIONS

There are many documented applications of remote sensing to groundwater hydrology. In general the photogrammetric interpretation of aerial photography and satellite imagery is the most prevalent in the literature. However, there are cases where analysts have taken advantage of some of the less common wavelengths available from modern remote sensing to solve specific problems. The following examples will illustrate current applications of all types of remote sensing data.

Shallow aquifers

Alluvial terrace deposits and glacial moraines are favourable sites for groundwater storage. These terrace deposits can be mapped by their relation to land cover and drainage. Thus, the presence of ox-bow lakes and old river channels are indicative of alluvial deposits (Farnsworth *et al.* 1984). Zigich and Kolm (1982) demonstrated how Landsat data provided a rapid and economical procedure for locating preglacial valleys in eastern South Dakota. Landsat bands 5 and 7 imagery taken in the dormant season were used with false-colour mosaics to delineate curvilinear patterns, drainage patterns and tonal patterns to infer the presence of buried valleys.

Shallow groundwater can be inferred in soil by changes in vegetation types and patterns (Nefedov and Popova 1972). Cultural features such as farming practices can be used to infer aquifers. Myers and Moore (1972), in a study of the James River lowland, showed that grains are the predominant crops in marginal areas, whereas pasture and alfalfa are the major land use on terraces and floodplains. In a more recent study, Rahn and Moore (1981) mapped surficial glacial aquifers using temporal Landsat imagery. Their interpretation was based on spectral anomalies associated with different cropping patterns. Groundwater recharge and discharge areas within drainage basins can be inferred from soils, vegetation and shallow or perched groundwater.

Thermal data may also be used to identify alluvial deposits, shallow groundwater and springs and seeps through the differences in temperature resulting from near-surface groundwater. Myers and Moore (1972) found that thermal anomalies could be related to shallow groundwater in the Sioux basin. Heilman and Moore (1981) discuss the HCMM (Heat Capacity Mapping Mission) thermal infrared data taken at appropriate times of the diurnal and seasonal temperature cycles and how these can be used to infer shallow groundwater. However, these authors point out that vegetative cover with differing transpiration rates, topography and other environmental factors confuses the interpretation. In a later paper, Heilman and Moore (1982) were able to obtain significant correlations between empirically estimated surface temperatures from the HCMM data and groundwater depths.

The relationship between groundwater depth and surface temperature was studied by van de Griend et al. (1985). The results of this modelling study for a

---

Figure 8.1 Identical areas on the sediment blanket of the eastern Sahara (Schaber et al. 1986). (a) SIR-A image, (b) Landsat MSS band 7 image. A (dark radar response), areas of confluence of several ancient river systems; B (bright response), grazing, stream-cut outcrops of iron-rich 'Nubia sandstone' and quartzitic sandstone mantled by thin eolian sand deposits; C (bright response), terrace or divide mantled by thin sand, isolated outcrops and rubble of 'Nubia' rocks, sinuous wadi (dark) of possible Pleistocene age (broad arrows); D exposures of dense groundwater-deposited calcium carbonate (calcrete) within wide 'radar-river' channel; E (intermediate mottled response), isolated grazing outcrops and blocky rubble of 'Nubia sandstone'; F (dark mottled response), low denuded interfluves of wide 'radar-river' valley, densely cemented below 20–50 cm depth by $CaCO_3$ forming radar interface; G (weak SIR-A return), extensive train of north-trending dunes 10 to 20 m high not penetrated by SIR-A; H (in (b) only) longitudinal dunes 2–3 m high that were penetrated by SIR-A signals.

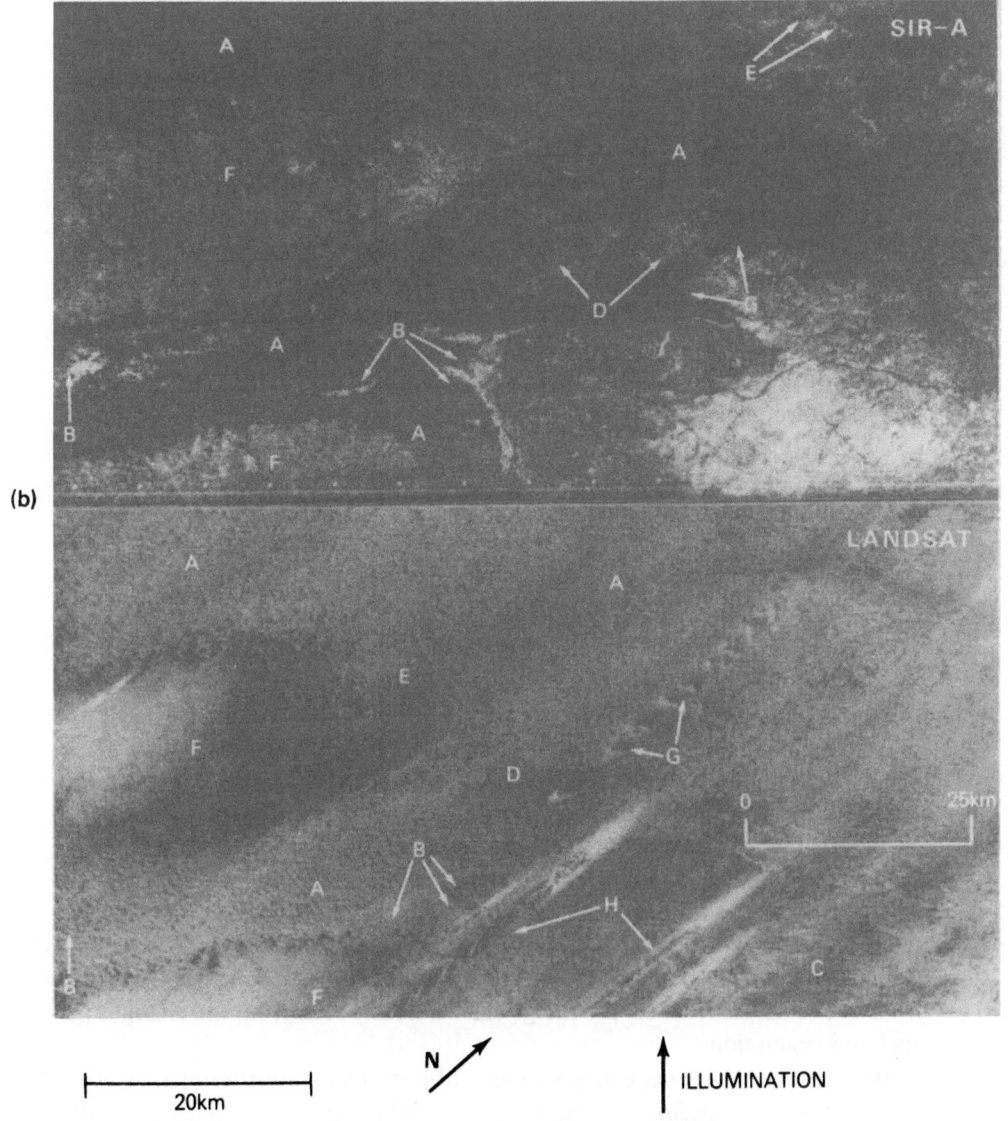

(a)

SIR-A

(b)

LANDSAT

0   25km

20km

N

ILLUMINATION

sandy loam showed that groundwater depths shallower than 90 cm are detectable but deeper groundwater is not. These results must be qualified by the particular conditions modelled using a coupled soil moisture and temperature model (Camillo *et al*. 1983). However, without this type of model it is difficult to isolate the effects of a water table from relief and meteorological effects.

The potential of SAR data for shallow groundwater exploration in arid and hyperarid regions has been dramatically demonstrated with SIR-A and SIR-B imagery. The penetration capability of long-wave radar and its importance to arid lands geology was illustrated by the SIR-A data taken over the eastern Sahara (McCauley *et al*. 1982, 1986). The SIR-A imagery from this region, one of the driest on the Earth, revealed a vast, previously uncharted network of valleys and smaller channels buried by the desert sands. This drainage network has been obscured from visible sensors as demonstrated in Figure 8.1 which compares a Landsat scene with the SIR-A image (Schaber *et al*. 1986).

Soil moisture is another clue that may indicate the presence of shallow groundwater. The measurement of soil moisture by remote sensing techniques was covered in Chapter 7.

Solid geology

Major water-bearing areas in solid rock formations are found along fault and fracture traces because the hydraulic conductivity and potential storage are both greatest in these areas. Surface expressions of fault and fracture patterns are often visible on aerial photographs and satellite images as linear or curvilinear features. Of course not all linear features are faults or fractures; consequently, contextual information must be used to select the most likely linear features for confirmation on the ground. Schowengerdt *et al*. (1981) demonstrated how photolineaments mapped using computer-enhanced Landsat imagery could be used to identify areas for more detailed mapping at aerial photography scales. Figures 8.2 and 8.3 are examples of the contrast stretching of Landsat bands 5 and 7 for a region of north-eastern Arizona where the faults marked on the image can be clearly seen from the changes in relief and vegetation.

Attempts have been made to reduce the subjectivity of interpreting imagery. Spatial filtering to define lineaments using global operators (Schowengerdt *et al*. 1981) and specialized operators such as Hough transforms (Gurney and Williamson 1981; Wadge and Cross 1989) have been attempted. Unfortunately, these have met with only partial success because of the complex relationships between terrain and the solid geological structure. Spatially filtered images may exhibit spurious features that may be misinterpreted (Schowengerdt *et al*. 1981).

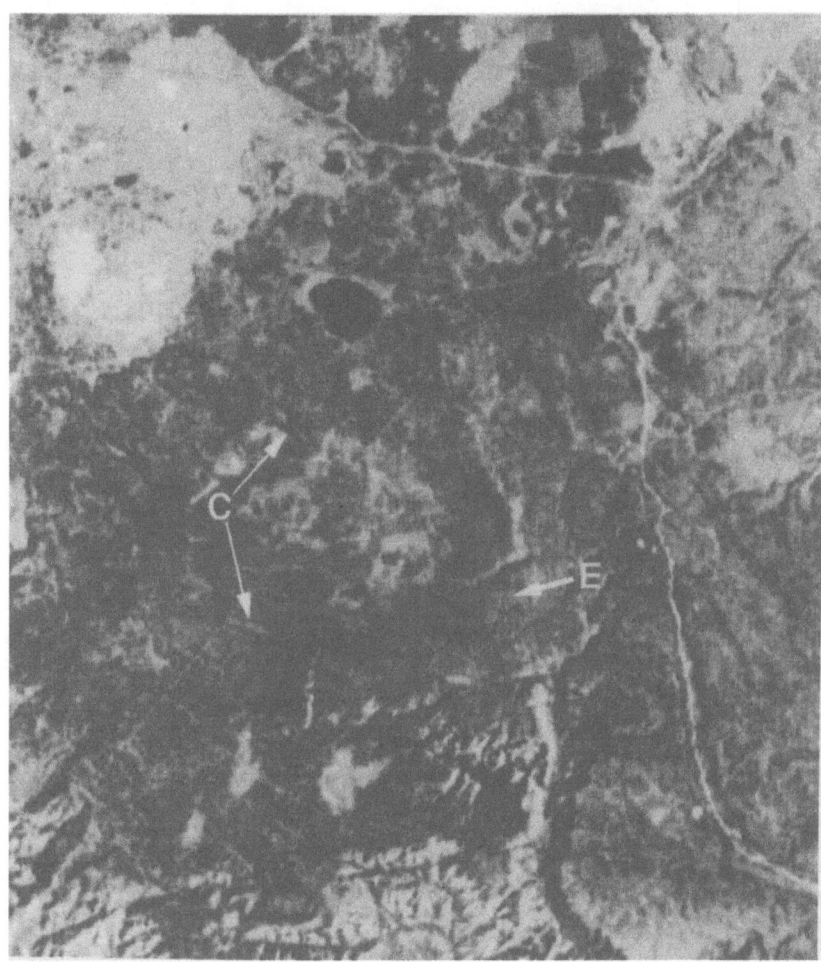

Figure 8.2 Band 5 contrast stretch of Landsat data in north-eastern Arizona, where faults can be clearly seen: site A (C, visible expression of north-west-trending faults; E, poorly expressed segment of Oak Creek fault) (Schowengerdt *et al*. 1981; reproduced with permission from The American Water Resources Association).

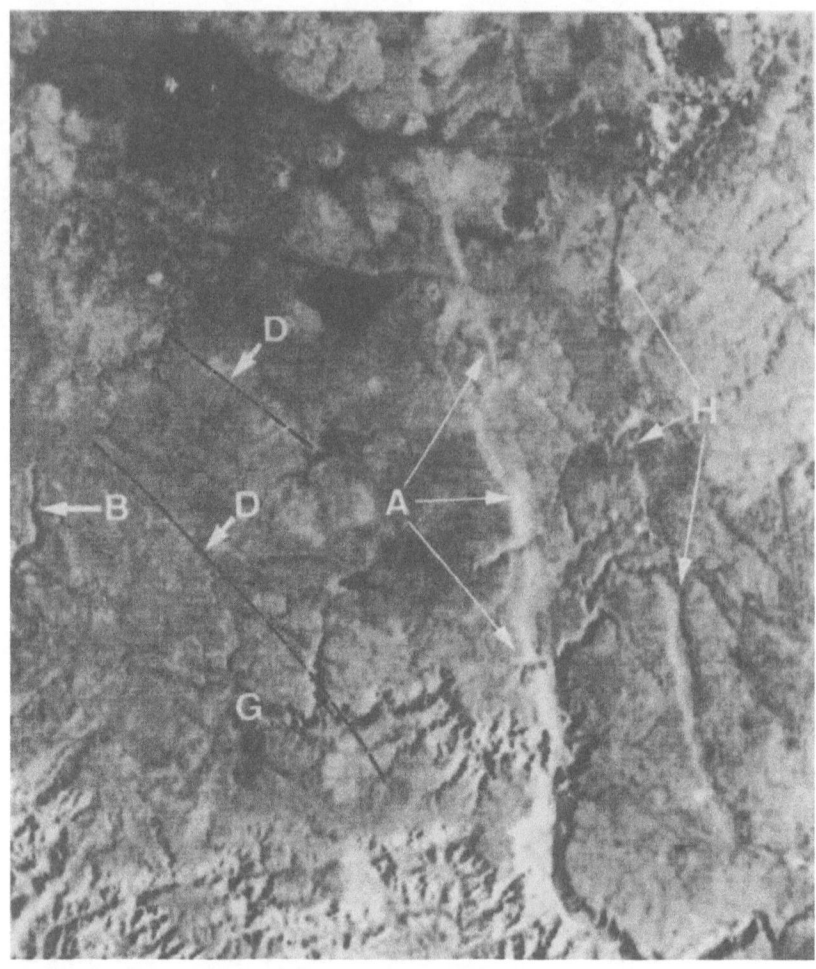

Figure 8.3 Band 7 contrast stretch of Landsat data in north-eastern Arizona, where faults can be clearly seen: site A (A, very good enhancement of Oak Creek fault; B, plot position and expression of vertical fault; D, plotted positions of large north-west-trending faults; G, enhanced expression of north-west-trending features; H, enhancement and plot of fault system just east of the Oak Creek system) (Schowengerdt *et al.* 1981; reproduced with permission from The American Water Resources Association).

Kaufman *et al.* (1986) demonstrated how high-resolution (Landsat TM) satellite data can be used to support and complement hydrogeological research. The objective of their research was to define the karst drainage patterns in the Peloponnesus area of Greece. A combination of TM bands 1, 4, and 7 was used in their analysis (see Plate 4) to identify different lithological units on the basis of texture and vegetation densities. Figure 8.4 shows the resulting geological units from the TM analysis and Figure 8.5 shows the resulting structural map derived from the TM image. Since the fresh-water drainage from the karst area is towards the Gulf of Argos, it was important to identify the major flow paths. The TM thermal band was able to be used to detect several of the submarine springs near the coast.

Hydrogeological interpretation of Landsat data has been shown to be a valuable survey tool in areas of the world where little geological and cartographic information exists. Gurney *et al.* (1982) demonstrated how satellite data can be used to assess groundwater resources in a poorly mapped area of the Kalahari Plateau in Botswana. Landsat data and limited aerial photography were used to find geologically significant features to aid a field team studying the underlying sandstone aquifer. Kruck (1981) also used interpretation of Landsat imagery to develop a regional representation of groundwater conditions in the Kalahari beds of northern Botswana. One of the concerns in this study was the salinity of the groundwater which was isolated by identifying an ancient lake that had since dried up and been covered by alluvial and aeolian deposits.

In another application of satellite data to groundwater exploration, Zall and Russell (1981) used digitally enhanced and photographically enlarged Landsat imagery. The digital processing enhanced the grey levels which represent subtle tonal changes caused by lithographic features, geological structure or soil moisture. The digitally processed imagery also enhances linear features, fractures, faults and drainage patterns that may be difficult to interpret with non-processed data. An example of the geological interpretation of the Bobo–Dioulasso area in Upper Volta, West Africa, is shown in Plate 5 and Figure 8.6.

Radar is very responsive to surface roughness and changes in topography that can be used to describe the lithography, land forms and past tectonic activity. The capability of radar to penetrate cloud cover and vegetation makes it doubly valuable as a reconnaissance tool for areas of the world that have previously been inaccessible to large-scale mapping and geological surveys. Figure 8.7 shows radar scenes from regions of Indonesia which are not visible to optical sensors because of the dense tropical rain forest and rainfall. These images and their interpretation offer proof of the utility of radar for geological mapping. Although radar cannot detect any but the most shallow groundwater directly, its use in detecting land forms and geological structure in inaccessible regions of the world make it a valuable tool for groundwater exploration.

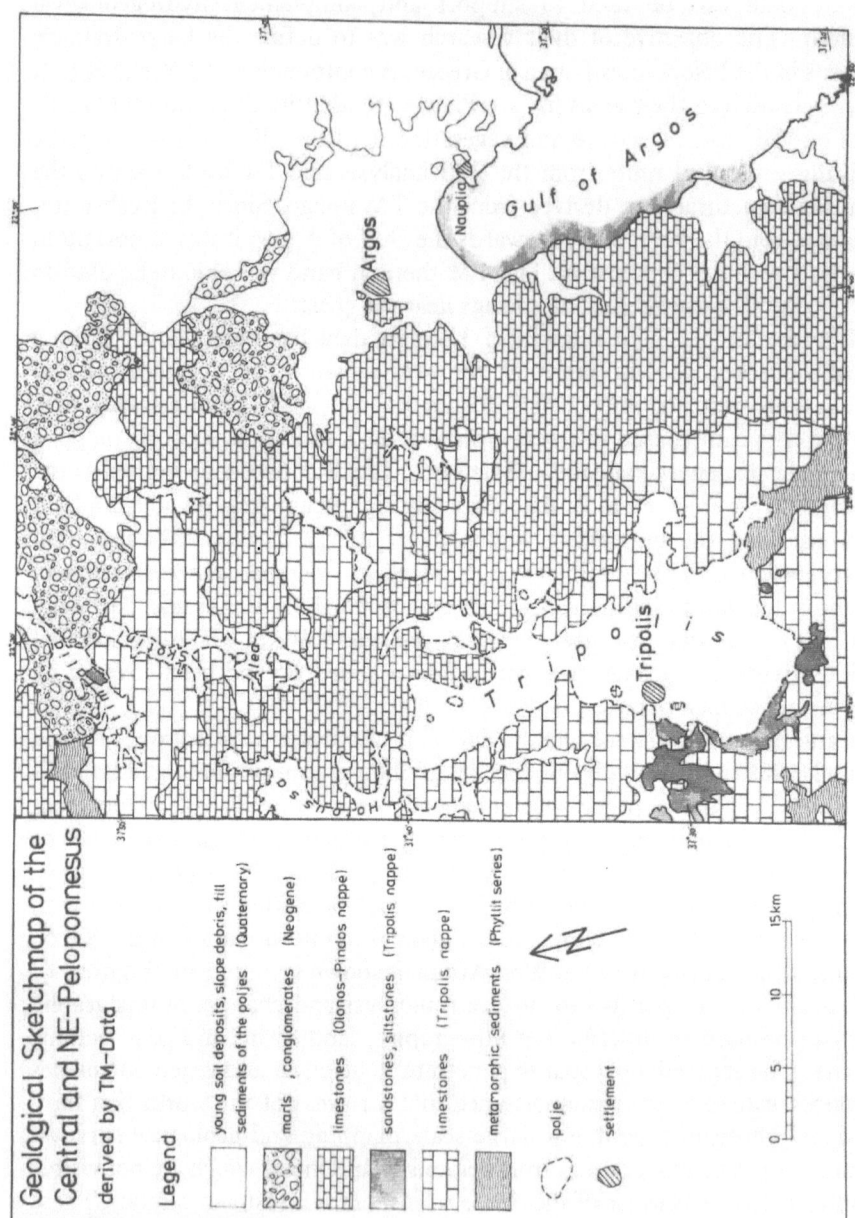

Figure 8.4 Geological units defined by Kaufmann *et al.* (1986) from **Plate 4.** (Reproduced with permission from IEEE.)

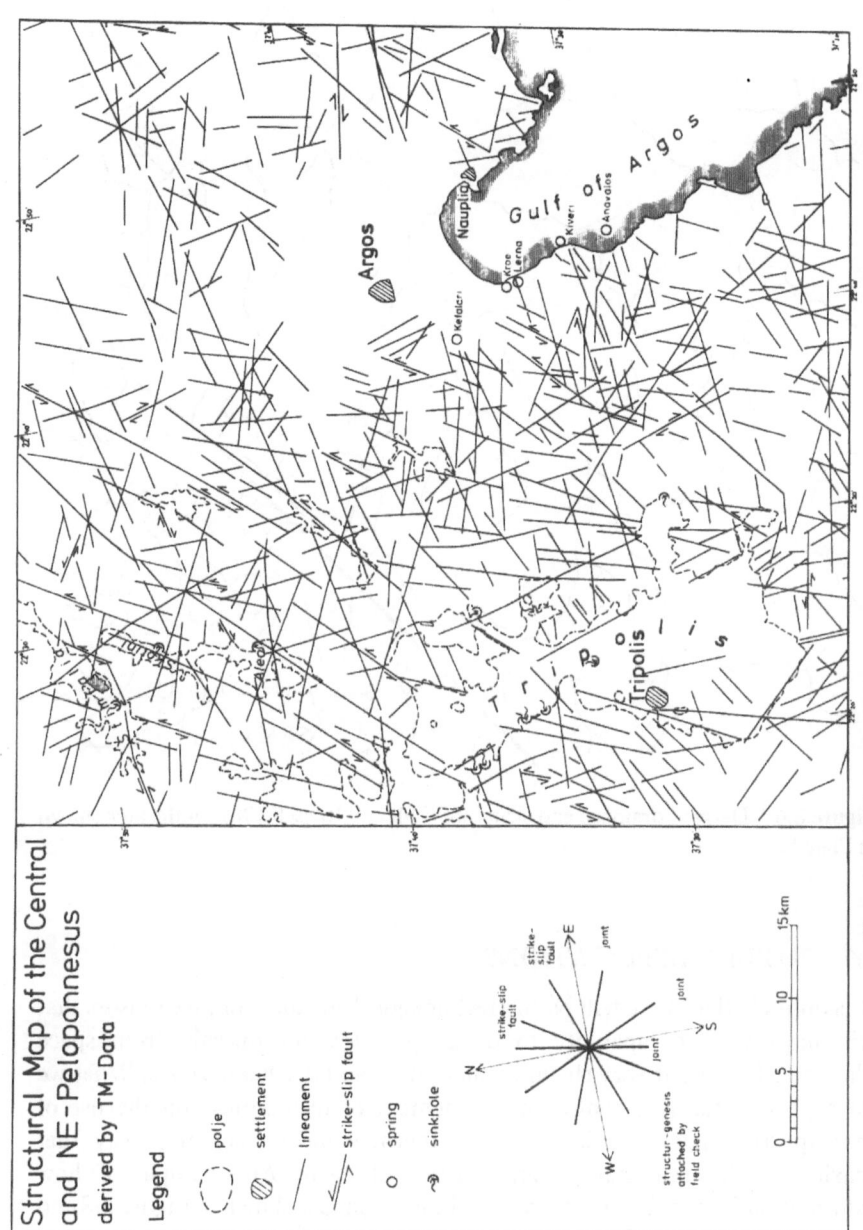

Figure 8.5 Geological structure map derived by Kaufmann *et al.* (1986) from Plate 4. (Reproduced with permission from IEEE.)

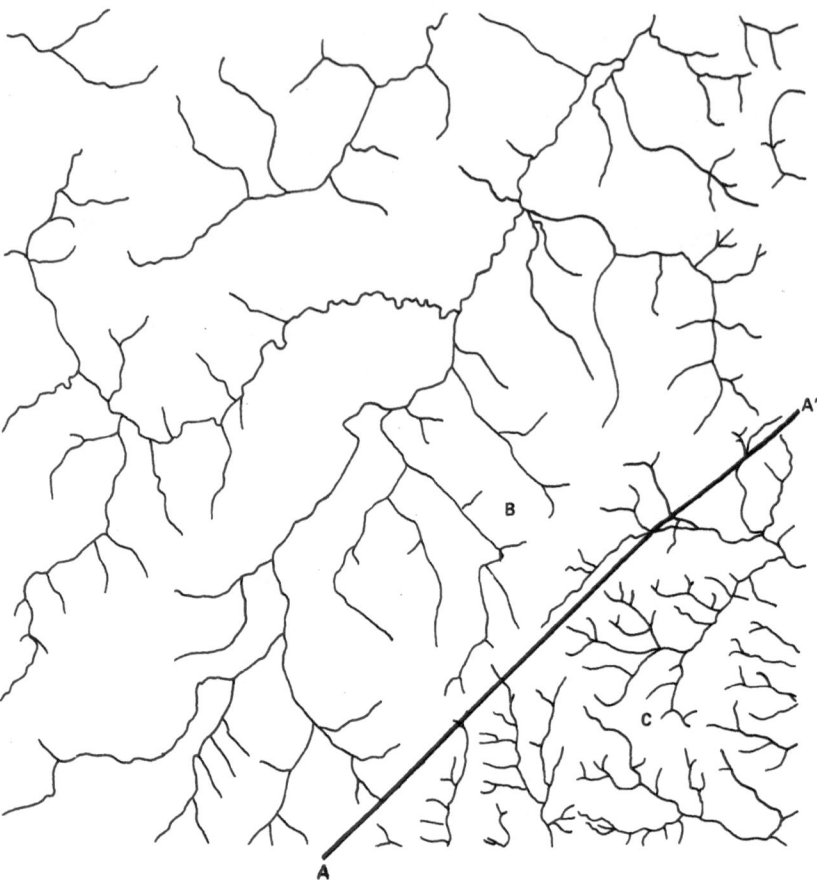

Figure 8.6 Derived drainage and structure (Zall and Russell 1981) of the area shown in Plate 5.

8.7 FUTURE APPLICATIONS

It is unlikely that there will be any technological advances or new sensors that will improve our capability to detect groundwater directly from space platforms for all but very shallow water tables and discharge zones. What we do expect to see is a continuing refinement of techniques and the use of multispectral data (visible, infrared, thermal and radar) in a way that maximizes the important characteristics of each. An example of how information from different sensors can be combined is shown in Figure 8.8 and Plate 6 (Chavez and Sanchez 1981). It is important to note how the most important features of each are maintained and that combined they produce a very vivid and detailed image.

Airborne exploration for groundwater has recently been conducted using electromagnetic prospecting sensors developed for the mineral industry. This type of equipment has been used to map aquifers at depths greater than 200 m (Paterson and Bosschart 1987). We can probably expect further advances in this area of remote sensing, particularly through the improvement of wide-band sensors and computerized interpretation techniques.

REFERENCES

Camillo, P. J., Gurney, R. J. and Schmugge, T. J. (1983) A soil and atmospheric boundary layer model for evapotranspiration and soil moisture studies. *Water Resour. Res.* **19**, 371–80.

Chavez, P. S., Berlin, G. L. and Tarabzouni, M. A. (1983) Discriminating lithologies and surficial deposits in the Al Hisma plateau region of Saudi Arabia with digitally combined Landsat MSS and SIR-A images. *Proc. Natl. Conf. on Resource Management Applications* **4**, 22–34.

Chavez, P. S. and Sanchez, T. (1981) Digital combination of Landsat and Seasat images. *Eos* **62**, 294.

Farnsworth, R. K., Barrett, E. C. and Dhanju, M. S. (1984) Application of remote sensing to hydrology including groundwater. *Unesco IHP Project* A. 1.5, p. 122.

Gurney, C. M. and Williamson, K. (1981) Use of Landsat data in detecting tectonic features in Central America. *IGS Rep.* Wallingford, England.

Gurney, R. J., Debney, A. G. P. and Gordon, M. R. (1982) The use of Landsat data in a groundwater study in Botswana. *J. Appl. Photogr. Eng.* **8**, 138–41, 232.

Heilman, J. L. and Moore, D. G. (1981) Groundwater applications of the Heat Capacity Mapping Mission. *Satellite Hydrology*, American Water Resources Association, Minneapolis, MN, pp. 446–9.

Heilman, J. L. and Moore, D. G. (1982) Evaluating depth to shallow groundwater using Heat Capacity Mapping Mission (HCMM) data. *Photogram. Eng. Remote Sensing* **48**, 1903–6.

Kaufman, H., Reichart, B. and Hotzl, H. (1986) Hydrogeological research in Peloponnesus karst area by support and completion of Landsat-thematic data. *Proc. IGARSS'86 Symp., Zürich, 8–11 September 1986*, ESA SP-254, 1, pp. 437–41.

Kruck, W. (1981) Hydrologic interpretations of Landsat imagery in arid zones of South Africa and West Africa. *Satellite Hydrology*, American Water Resources Association, Minneapolis, MN, pp. 408–15.

McCauley, J. F., Breed, C. S., Schaber, G. G., McHugh, W. P., Haynes, C. V., Issaw, B., Grolier, M. J. and El-Kilani, A. (1986) Paleodrainages of the eastern Sahara — The Radar Rivers revisited. *IEEE Trans. Geosci. Remote Sensing* **GE-24**, 624–48.

McCauley, J. M., Schaber, G. G., Breed, C. S., Grolier, M. J., Haynes, C. V., Issawi, B., Elachi, C. and Blom, R. (1982) Subsurface valleys and geoarchaeology of the eastern Sahara revealed by Shuttle radar. *Science* **218**, 1004–20.

Moore, G. K. (1979) Prospect for groundwater with Landsat images. *Proc. Semin. on Remote Sensing Applications and Technology Transfer for International Development*, Environmental Research Institute of Michigan.

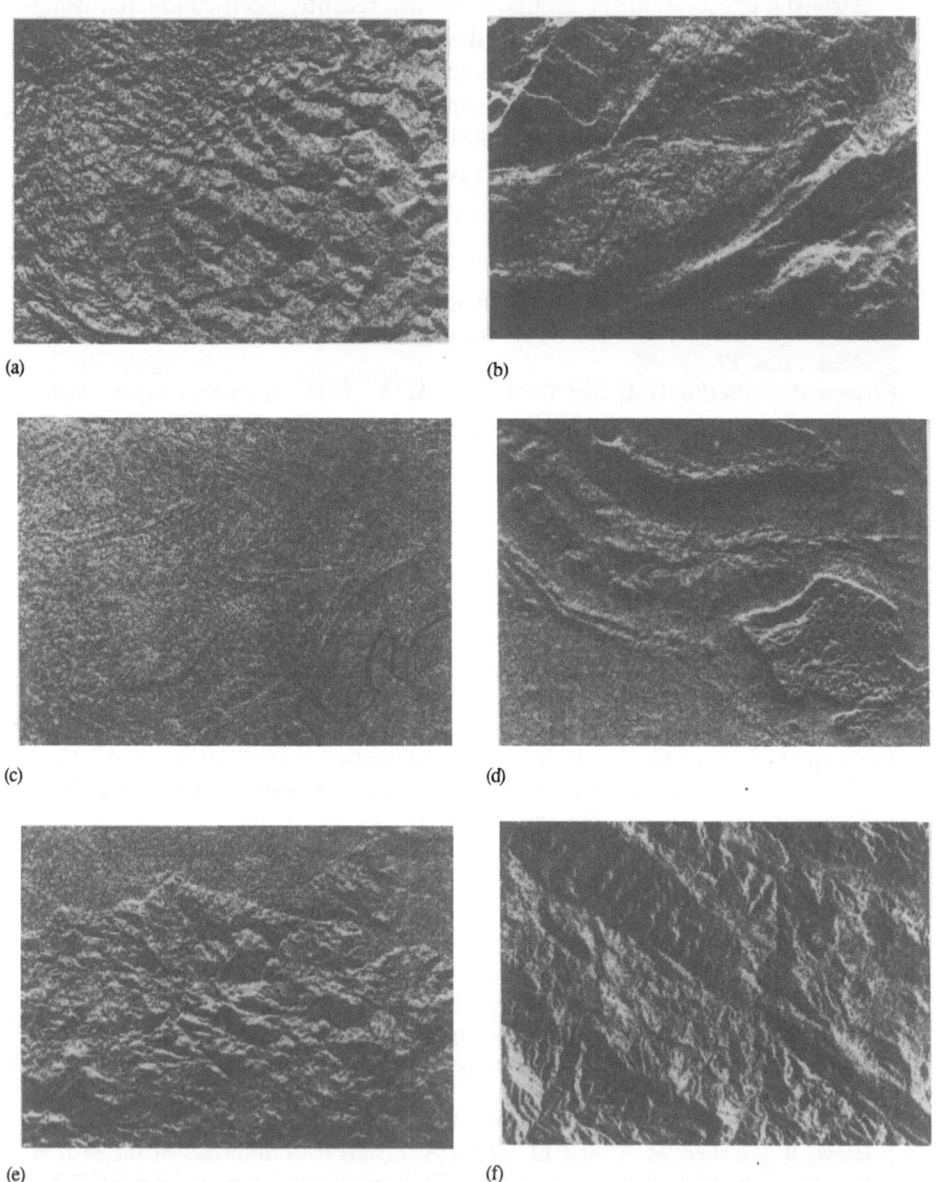

(a)

(b)

(c)

(d)

(e)

(f)

Figure 8.7 (a–l) SIR-A images of Indonesian structural features. Each image is 28 km wide. Interpretation maps of Indonesian structural features are also shown (Sabins 1983; reproduced with permission.)

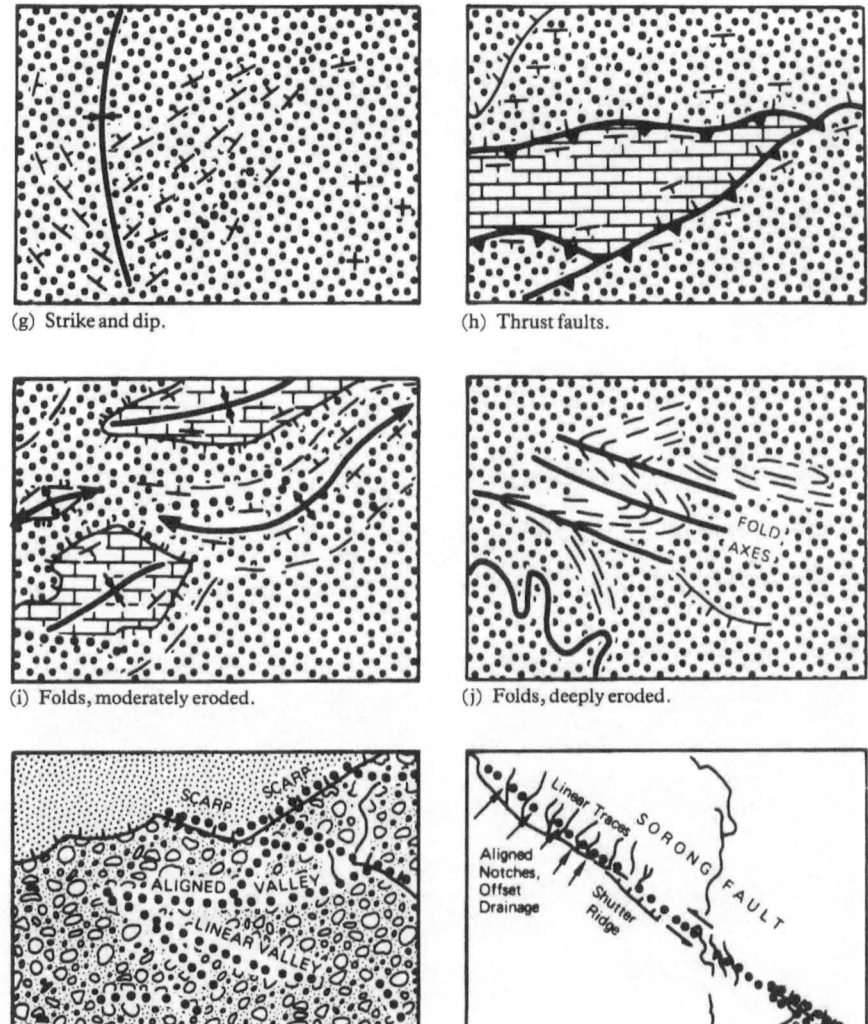

(g) Strike and dip.

(h) Thrust faults.

(i) Folds, moderately eroded.

(j) Folds, deeply eroded.

(k) Lineaments.

(l) Strike–slip fault.

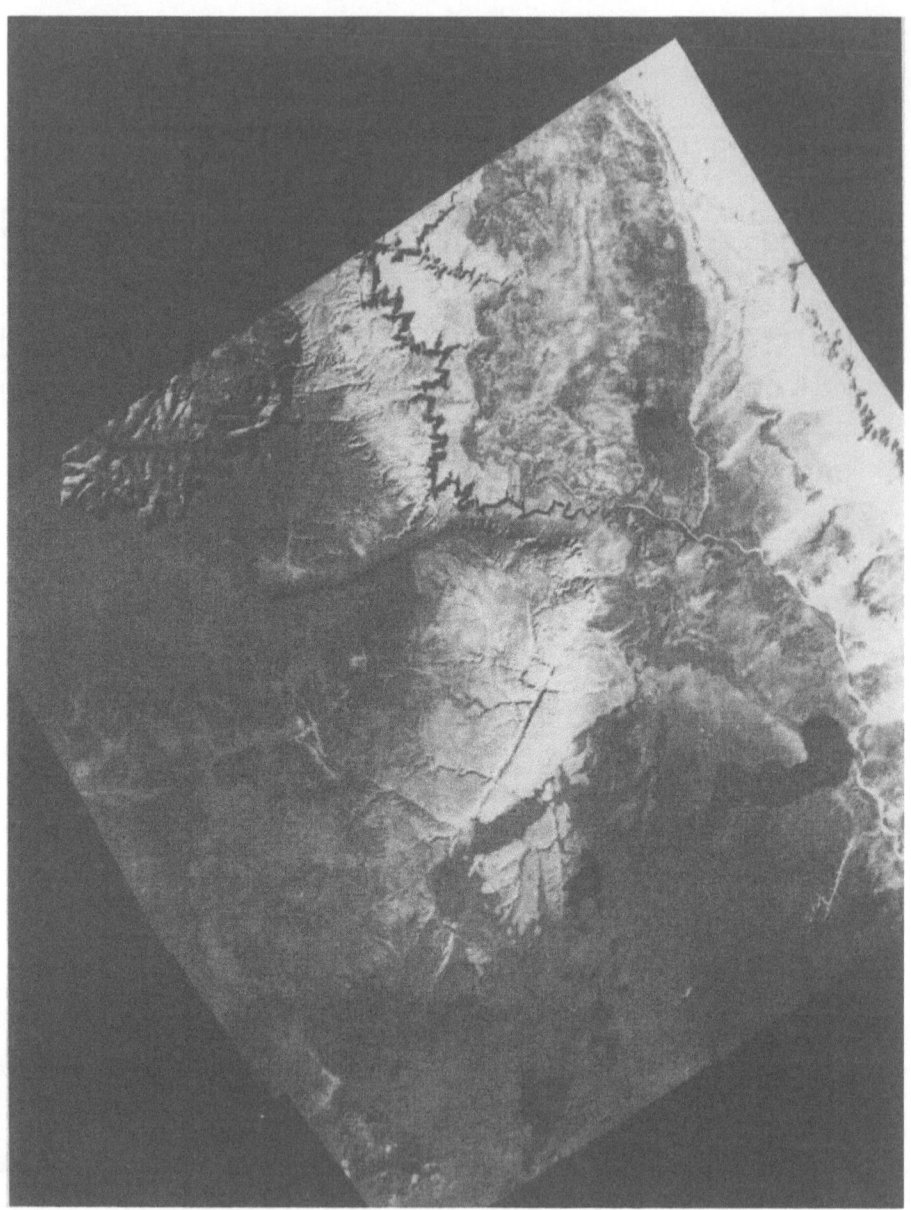

a)

Figure 8.8   (a) Landsat (bands 4, 5 and 7) image of San Francisco volcanic field north of Flagstaff, Arizona. (b) Seasat SAR image of same area. See also Plate 6. (Chavey and Sanchez 1981).

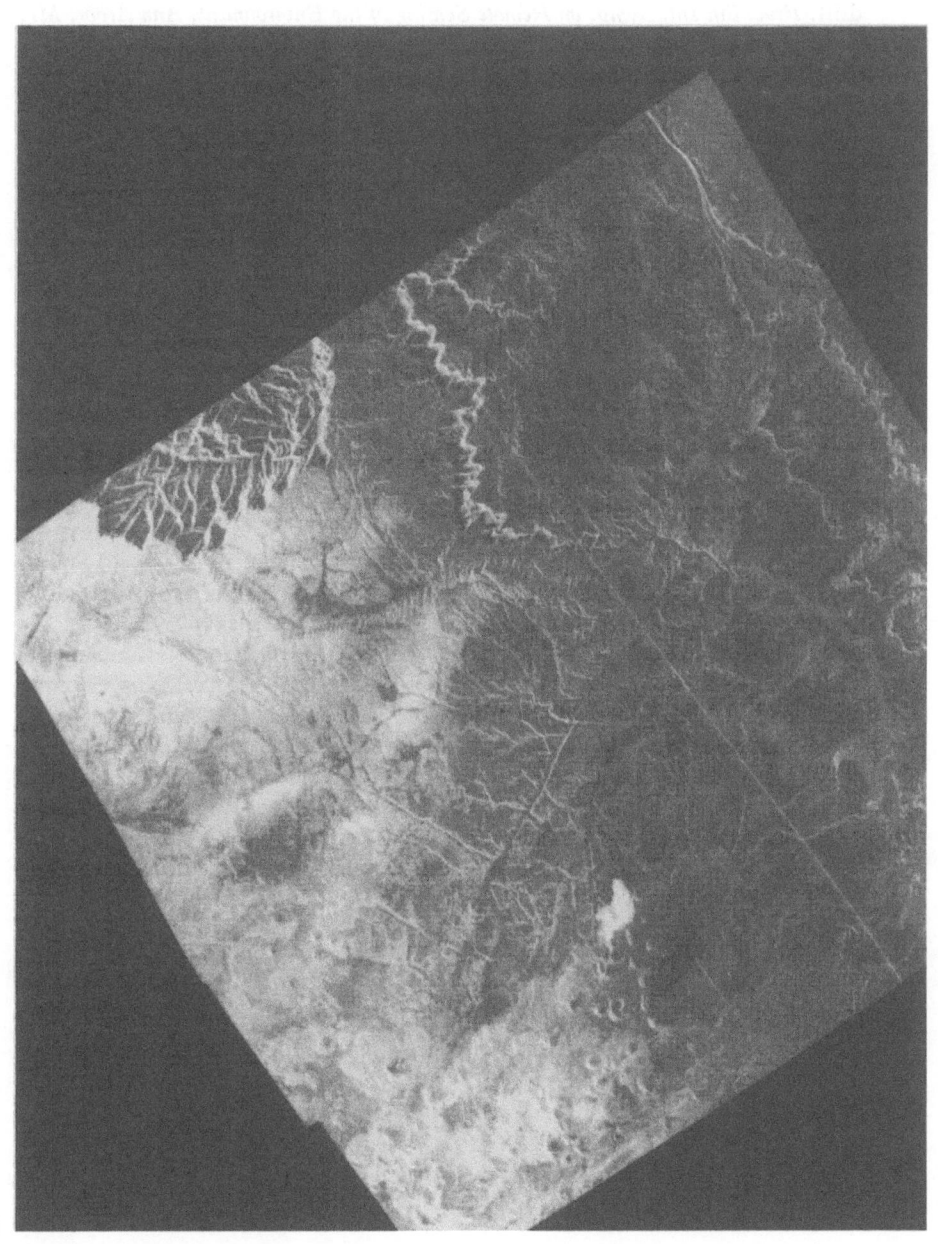

Myers, V. I. and Moore, D. G. (1972) Remote sensing for defining aquifers in glacial drift. *Proc. 8th Int. Symp. on Remote Sensing of the Environment, Ann Arbor, MI*, Environmental Research Institute of Michigan, Ann Arbor, MI, pp. 715–28.

Nefedov, K. E. and Popova, T. A. (1972) *Deciphering of Groundwater from Aerial Photographs*, Amerind, New Delhi, transl. (1969) *Deshifrirovanie Gruntovykh vod po Aerofotosnimkam*, Gidrometeorologicheskoe Press, Leningrad.

Paterson, N. R. and Bosschart, R. A. (1987) Airborne geophysical exploration for groundwater. *Groundwater* 25, 41–50.

Rahn, P. H. and Moore, D. G. (1981) Landsat data for locating shallow glacial aquifers in eastern South Dakota. *Satellite Hydrology*, American Water Resources Association, Minneapolis, MN, pp. 398–406.

Ray, R. G. (1960) Aerial photographs in geologic interpretation and mapping. *USGS Prof. Pap.* 373, pp. 230.

Revzon, A. L., Burleshin, M. I., Krapilskaya, N. M., Sadov, A. V., Svitneva, T. V. and Semina, N. S. (1983) Study of the desert geological environment with the aid of aerial and space imagery. *All-Union Sci. Res. Inst. Hydrol. Eng. Geol.*, Moscow, USSR.

Sabins, F. F., Jr (1983) Geologic interpretation of space shuttle radar images of Indonesia. *Am. Assoc. Petrol Geol. Bull.* 67, 2076.

Schaber, G. G., McCauley, J. F., Breed, C. S. and Olhoeft, G. R. (1986) Shuttle imaging radar: physical controls on signal penetration and subsurface backscatter in the Eastern Sahara. *IEEE Trans. Geosci. Remote Sensing* GE-24, 603–23.

Schowengerdt, R., Babcock, E. M., Ethridge, L. and Glass, C. E. (1981) Correlation of geologic structure inferred from computer enhanced Landsat imagery with underground water supplies in Arizona. *Satellite Hydrology*, American Water Resources Association, Minneapolis, MN, pp. 387–497.

van de Griend, A. A., Camillo, P. J. and Gurney, R. J. (1985) Discrimination of soil physical parameters, thermal inertia and soil moisture from diurnal surface temperature fluctuations. *Water Resour. Res.* 21, 997–1009.

Wadge, G. and Cross, A. M. (1989) Identification and analysis of the alignments of point-like features in remotely-sensed imagery: Volcanic cones in the Pinacate Volcanic Field, Mexico. *Int. J. Remote Sensing* 10, 455–74.

Zall, L. and Russell, O. (1981) Ground water exploration programs in Africa. *Satellite Hydrology*, American Water Resources Association, Minneapolis, MN, pp. 416–25.

Zigich, D. K. and Kolm, K. E. (1982) Evaluating the effectiveness of Landsat data as a tool for locating buried pre-glacial valleys in eastern South Dakota. *Photogram. Eng. Remote Sensing* 48, 1891–901.

9
Water quality

9.1 INTRODUCTION

Water quality is the general term that describes whether or not water is usable or whether or not the surrounding environment may be endangered by pollutants in the water. From a historical perspective, poor water quality was not a perceived problem before the industrial revolution and the rapid growth of cities. Early concerns with poor water quality focused on health and sanitation. Since about the turn of the century, attempts at water quality control have concentrated on sewage and industrial discharges. This approach involved very little mystery because most of these discharges came through a pipe or open channel. Treatment was a matter of commitment and economics. More recently, non-point source pollution has been the subject of both general concern and scientific investigation. Non-point source pollution is generally considered as part of storm water runoff. Urban, industrial and rural areas can all be major sources of non-point source pollution. A major problem associated with non-point pollution is identifying its source and eventual sink. This is not as simple as in the case of municipal sewage because there is no collection pipe. The sources of pollution are as varied as the sources of storm runoff, and as such they are unpredictable both in time and amount.

Remote sensing has an important role in water quality evaluation and management strategy. Sources of pollution are often easy to identify, especially when there are pipes or open channels discharging into a lake or river. Non-point source pollution can perhaps be evaluated best by remote sensing, especially when the spatially distributed nature of non-point source pollution is considered. The synoptic view provided by remote sensing gives an environmental scientist very different data from that which can be obtained with surface data collection and sampling. What may not be achieved with absolute accuracy is more than made up for by the spatial and temporal nature of the data. Monitoring large areas on a frequent basis can only be achieved economically with remote sensing. Unfortunately, remote sensing is pretty much limited to surface measurements of turbidity, suspended sediment, chlorophyll, eutrophication and temperature. However, these characteristics of water quality can be used as indicators of more specific pollution problems. It should be emphasized that remote sensing is seldom the only approach to monitoring. Effective use of remote sensing measurements can only be made when they are used as ancillary data with other data or information.

9.2   GENERAL APPROACH

The region of the electromagnetic spectrum that includes visible and infrared light is useful for detecting indicators of water quality. Measurements in this part of the spectrum utilize reflected electromagnetic energy. Thermal infrared is also used for measuring water quality but it uses a direct measure of the emitted energy. The microwave region is not particularly useful for determining indicators of water quality because there is little if any penetration into the water. Detection of oil slicks or other surface contamination is the exception to this general statement. The approach differs depending on whether the reflected or the emitted energy is being measured.

In general, a relationship between a given water quality parameter and the reflectance must be established. Because the intensity and colour of light is modified by the volume of water and its contaminants, an empirical relationship can be established between the reflectance measurement and the water quality sample. The degree to which light is attenuated by water varies with the wavelength of the light and with the nature, concentration and colour of the water quality contaminant. Figure 9.1 shows how light is attenuated by clear water. Note that there is practically no penetration of the infrared energy into the water whereas approximately 60% of the blue light ($\simeq 0.5 \ \mu$m) is still measurable at 20 m.

The presence of sediment in water changes the backscattering characteristics of the water dramatically. Figure 9.2 illustrates how the sediment concentration increases the reflected energy compared with the reflected energy from clear water. Note also how the peak of the curves shifts to the longer wavelengths.

The empirical nature of this approach limits its usefulness because a unique relationship must be developed for each environmental situation. Even after an empirical relationship has been established, it is likely to change because the type of constituent in the water may not remain constant. Also, the angular relationship between the Sun and the sensor will change with the time of year

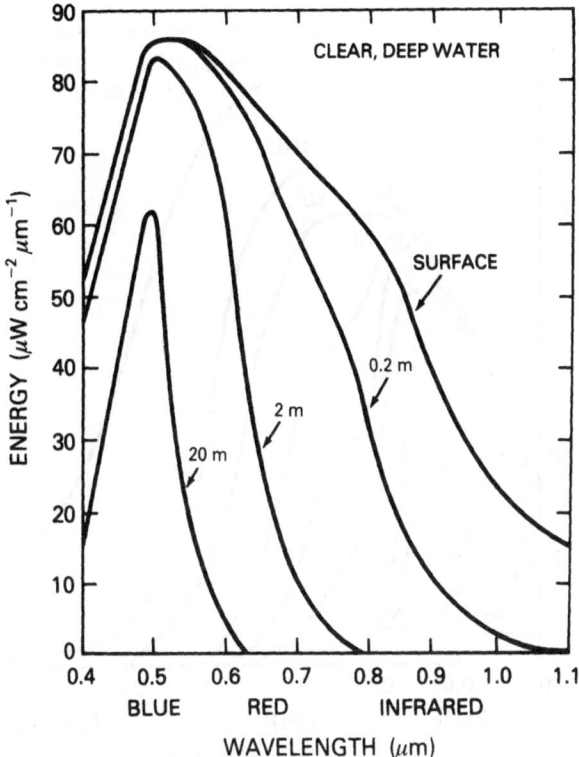

Figure 9.1 An illustration of how light is attenuated by clear water. Note that in the infrared region there is very little penetration into the water.

and the composition of the atmosphere will not remain constant. The latter two effects can be compensated for, but there is no way to measure the temporal changes in the water quality constituents directly with a high degree of confidence. Periodic field sampling programmes must therefore be carried out to verify the empirical relationships, perhaps every year or several years.

Applications of thermal infrared remote sensing for water quality generally takes advantage of the 8–14 μm waveband. Not only is this region of the spectrum an efficient atmospheric window, but it also contains the region of the maximum radiant emittance for most Earth features. Water behaves much like a black body with its peak emission occurring between 9 and 10 μm and with an emissivity close to unity. In general, most water quality monitoring can assume constant emissivities in the 8–14 μm range, but it is important for quantitative work to know of the variability that occurs over time and with wavelength.

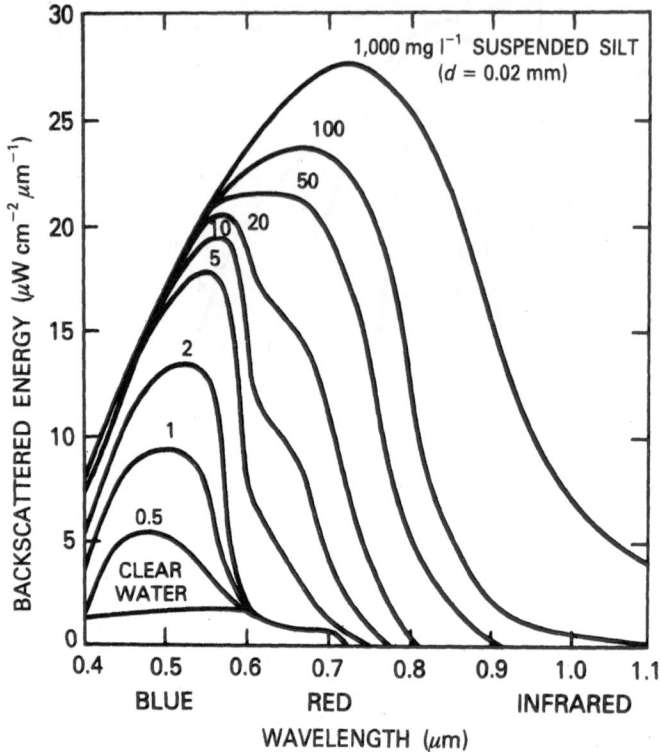

Figure 9.2 An illustration of the increase in backscattered energy as the sediment concentration increases.

Water quality monitoring with thermal infrared is based on measuring temperature differences spatially. The interpretation of thermal data generally relies on some ancillary information, such as knowledge of the location of a discharge pipe, to infer a water quality impact. However, thermal pollution at the water surface can be measured directly with thermal scanners that can be used to derive a very detailed map of actual temperature.

9.3 DETERMINING WATER QUALITY

Remote sensing can only measure reflected or emitted energy from the surface or near surface of water. Remote sensing applications are limited to those characteristics that can be observed, and so there is a limit on the water quality characteristics that can be detected. As discussed above, an empirical relationship between the water quality parameter and one or more spectral bands must be established to infer the water quality status of a water body. Water quality indicators such as colour, turbidity, chlorophyll and suspended solids have been successfully used in many applications. These parameters and some of the approaches used to evaluate them are discussed in the following subsections.

Colour

Colour is a non-specific water quality parameter that describes qualitative information about the biological productivity and the general chemical make-up of water bodies. True colour (specific colour) is determined by substances in colloidal suspension or in solution. Apparent colour, on the other hand, is usually what is observed and is caused by light reflecting on suspended materials, the bottom, or reflected sky. To determine true colour, *in situ* sampling must be performed.

Observations of colour are often made to evaluate the amounts of living and non-living substances in the water. Plankton algae are commonly associated with specific colours: blue–green algae are linked to dark-green colours; diatoms impart a yellow or yellow–brown hue; some zooplanktons impart reddish hues; whereas humus can cause water colours to vary from green to dark brown. Dissolved chemicals can impart a specific colour to a water solution, but frequently these changes are too subtle or are obliterated by other sources of colour in natural water and thus are generally not detectable by common remote sensing techniques. Inexact corrections for atmospheric absorption of reflected light also inhibit the measurement of true colour by remote sensing techniques.

Existing satellites can provide colour data useful for determining water quality indicators. The Nimbus-7 coastal zone scanner and the Landsat and SPOT series can provide valuable spatial and temporal maps of specific colours that can be related to specific water quality indicators. Feldman *et al.* (1984) used Nimbus-7 data to track changes in colour that they associated with phytoplankton concentrations. These data documented a major redistribution of phytoplankton around the Galapogos Islands. The authors associated these changes in food resources with the reproductive failure of sea birds and marine mammals in the 1982–1983 El Nino.

Viollier *et al.* (1978) performed a series of ocean colour measurements with a special airborne radiometer to estimate the chlorophyll content over the Gulf of Guinea. Fourteen low-level flights provided data which, when mapped, showed the detailed structure of boundaries due to upwelling. Chlorophyll content was found to affect the albedos at the two wavelength bands of 466 and 525 μm, whereas the albedos of longer wavelengths were found to be affected primarily by light-scattering particles (presumably sediments).

Turbidity

Turbidity is another non-specific water quality indicator that is used to characterize receiving waters. Turbidity is defined as optical transparency or the degree of opaqueness produced in water by suspended particulate matter. There are several ways to measure turbidity, all of which are related to transparency. The commonly used Secchi disk (Reid 1965) is easy to use but quite subjective. Secchi disk measurements involve lowering a 20 cm white disk into the water. The depths at which the disk disappears and then appears again as it is drawn back up are averaged. As a rule of thumb, the disk disappears at a depth of roughly the 5% level of transmission of sunlight. More precise measurements can be obtained with light transmission instruments such as photometers and transmissometers; however, these do not particularly lend themselves to field surveys. The Secchi disk, on the other hand, can be easily and rapidly deployed from a boat to measure the spatial distribution of suspended materials.

The impreciseness of turbidity as a water quality indicator is based on the light attenuation in water by suspended sediment and organic materials. Thus, any relationship between turbidity and reflected radiation will necessarily be site and contaminant specific. Turbidity is not a uniform parameter, either spatially or temporally. It will change with changes in inflowing discharge and with internal currents, thermal layering and the life cycles in the water. Remote sensing provides an excellent method for tracking these spatial and temporal changes. Much of the dynamics of a water body as well as its water quality can be inferred from turbidity data. Thus, while they are not a precise point measurement of water quality, remote sensing measurements of turbidity give environmentalists and managers valuable spatial and temporal data. In many cases, such data could not be obtained economically by frequent point samples.

Turbidity estimates from satellite data are useful for tracking the movement of water masses within large lakes. Abiodun (1976) was able to identify five distinct water masses in Lake Kainji on the River Niger. Later work by Abiodun and Adeniji (1978) demonstrated how spectral classification of the lake water with sequential Landsat data could be used to chart the movement of different water masses. Figure 9.3 shows the spatial distribution of turbidity for four dates. This approach is very useful for large lakes where frequent synoptic surveys by field crews are either impossible or impractical.

Landsat data of the Kenyan coast were used to analyse the coastal movement of sediments from the Tana and Sabaki Rivers. Brakel (1984) mapped the relative concentrations of sediments by tracing the perimeter of turbid features with Landsat bands 4, 5 and 6. The turbidity plumes were developed by rainy-season runoff, and their movements controlled by monsoon winds and ocean currents.

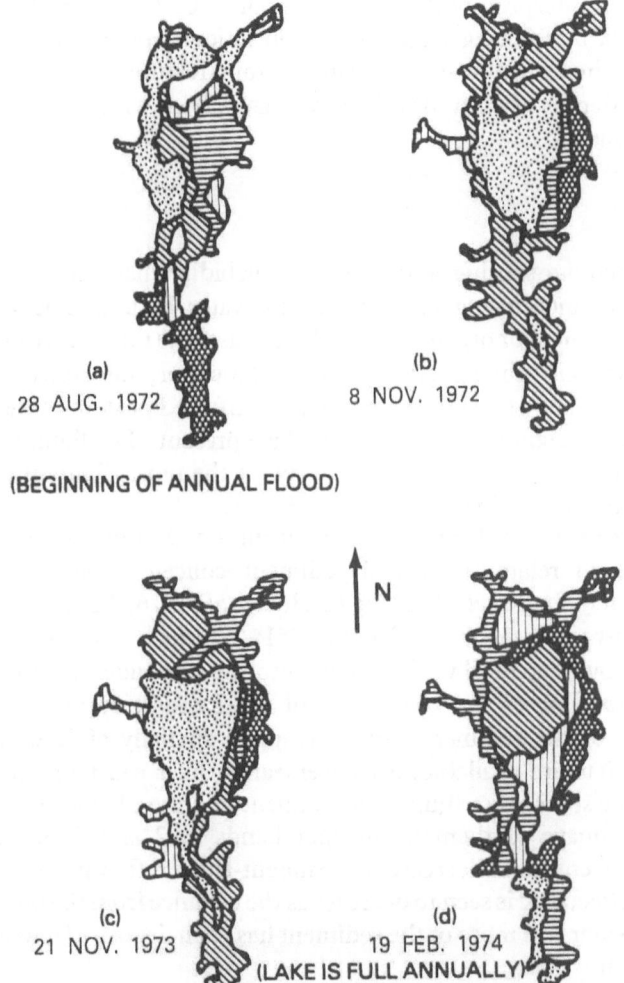

Figure 9.3 An illustration of the temporal changes of lake turbidity with season in Lake Kainji in Nigeria (after Abiodun and Adeniji 1978).

Multitemporal Landsat imagery has been used at different stages of the tidal cycle of aid in interpreting the bathymetry of a coastal inlet in New Zealand (Thomas 1980). Differing sediment concentrations from one date to the next enabled the investigator to delineate bottom features. The difference in penetration depths of each MSS band was associated with different sediment concentrations.

Sediment

A specific remote sensing application of turbidity measurements is in the detection and monitoring of sediment in water bodies. The presence of suspended sediment or organic materials increases the reflectance in the visible regions of the electromagnetic spectrum. However, the reflectance of the sediment and water remains low in the near-infrared portion of the spectrum unless there are significant amounts of algae present. The failure of a sensor to detect turbidity deeper than about 1 m, at the most, limits this approach to the detection of shallow suspended sediment.

A number of studies have been made using Landsat multispectral scanner (MSS) data to relate suspended sediment concentrations to reflectance measurements (Yarger et al. 1974; Ritchie et al. 1976; Munday and Alfoldi 1979; Aranuvachapum and LeBlond 1981; Whitlock et al. 1981). Landsat MSS data provide an ideal vehicle for monitoring suspended sediments in large water bodies. One of the advantages of the Landsat approach is that the distribution of the sediment can be mapped spatially and, when suitable calibration data are available, isoconcentration lines can be plotted. Plate 7 illustrates the spatial distribution of sediment in Lake Chicot, Arkansas. This false-colour image of thematic mapper bands 1, 2 and 3 shows how the sediment concentration decreases as sediment-laden inflow moves through the lake. The reflectance is seen to decrease as the distance from the point of inflow becomes greater and more of the sediment has been deposited and is no longer in suspension.

The temporal features of the Landsat data provide a time series of the sediment distribution and concentrations. In fact, we can even go back in time to generate a historical description of sedimentation activities in a given body of water. To do this the MSS data must be obtained for the period from 1972 (or thereabouts) to the present and the calibration data must be developed with current samples. In doing this the assumption must be made that the sediment characteristics have not changed with time, which may be considerable assumption. This is essentially the same assumption that must be made for using reflectance–concentration data for future monitoring of water bodies. However, for the future monitoring case, the calibration can be checked with additional sampling programmes at some later time.

The spatial resolution of Landsat MSS data limits the size of a water body

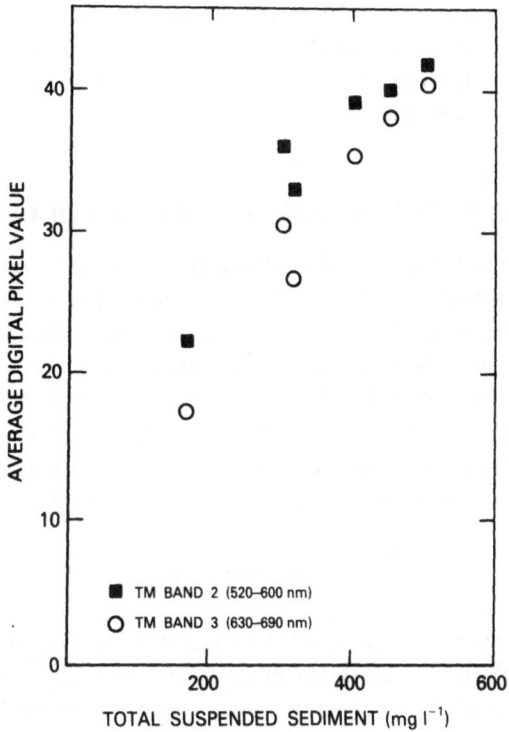

Figure 9.4 A relationship between the reflectance (average digital pixel value) and suspended sediment concentration for thematic mapper bands 2 and 3 (after Ritchie *et al.* 1984).

that can be monitored. The MSS pixel of about 80 m effectively limits the size of water impoundments to about 5 ha or more (Ritchie *et al.* 1984). These studies by Ritchie *et al.* using data from the thematic mapper (TM) aboard Landsat-4, have shown that such impoundments can be monitored effectively. The studies used TM data and suspended sediment samples taken from Lake Chicot, Arkansas. TM bands 1, 2, 3 and 4 all contained information related to near-surface suspended sediment with band 3 having the highest linear correlation between reflectance and sediment concentration. Figure 9.4 shows the relationship between reflectance and suspended sediment concentration for TM bands 2 and 3.

The increased spatial resolution available from the TM data expands the types of applications possible with remote sensing data. Ritchie *et al.* (1984) point out that this increased resolution can provide conservation agencies with a means of monitoring water quality in small agricultural impoundments and thus identifying watersheds with erosion problems. Water bodies that have high suspended sediment concentrations would pinpoint upstream watersheds where major erosion problems are present. Conservation agencies could use this information to plan erosion control strategies and focus their efforts on the most seriously affected areas.

9.4 EROSION AND NON-POINT SOURCE POLLUTION

Remote sensing has been used to estimate non-point source pollution for regional water quality studies and conservation plans. Non-point source pollution is closely related to land cover and use so that when coupled with soils and topography the potential water quality in runoff can be described. Several groups have used Landsat data for land cover input to their non-point source models (Schecter 1976; OKI 1975; Ragan and Rogers 1978). The OKI model estimates the erosion and sediment contribution from subareas. This approach is based on the evaluation of the universal soil-loss equation (Wischmeier and Smith 1978)

$$A = RKLSCP \qquad (9.1)$$

where A is the average annual soil loss per unit area, R the rainfall factor, K the soil erodability factor, L the slope length factor, S the slope steepness factor, C the cropping management factor and P the erosion control practice factor. For these studies the cropping management factor, C, and the erosion control practice factor, P, were lumped into one parameter and estimated from Landsat data.

A similar approach was used by an eight-county study group in central Indiana (Campbell 1979). In this study a non-point pollution potential (NPP) index was defined as

$$\text{NPP} = f\,(\text{soil characteristics} \times \text{agricultural land use intensity}$$
$$\times \text{proximity to water)} \qquad (9.2)$$

This index was applied to cells of approximately 62 acres (or 53 Landsat pixels). The agricultural land use intensity data were derived from the geocorrected Landsat data. The soil characteristics were determined from erodability data in soil surveys and the proximity to surface water was measured by a cell count from water bodies identified on topographic maps. A geographical information system was utilized to combine the data from the different data sets and assess the NPP index.

Another study by Pelletier (1985) describes a geographic information system (GIS) adaptation to the universal soil-loss equation (Wischmeier and Smith 1978) that uses data from either the Landsat MSS (79 m resolution) or the Landsat TM (30 m resolution) to determine the land cover, and subsequently the cropping management factor, c. Pelletier (1985) provides a good discussion of some of the accuracy limitations that may be encountered by automating the procedure with satellite and digital elevation data. Figure 9.5 is a diagram of the various data sources and data files managed by the GIS.

A quick-look model for estimating splash and sheet erosion on a regional scale has been developed by Mueksch (1984). This approach, based on empirical equations, uses slope areas and lengths delineated from Landsat imagery. The method has been tested in the East African coastal zone. Applications to other climatic and geographic areas should be carried out with caution; however, the method does demonstrate how remote sensing data may be used in areas where other data may not be available.

Colour-infrared aerial photography has been used to delineate homogeneous erosion mapping units, crop rotations, and management practices in a potato-farming area in New Brunswick, Canada (Stephens *et al.* 1985). This approach produced more accurate soil-loss estimates when combined with slope data taken from topographic maps and an improved soil erodability map. In addition, the use of aerial photographs allows a year-round analysis which had been limited to about 120 days by traditional methods.

In the cases cited above, soil erosion has not been measured directly by remote sensing techniques. In all cases, a surrogate of erosion related to land use has been determined. The spatial resolution available from common remote sensing products limits its use to this type of application and for most large-area surveys and inventories it is quite adequate. However, soil erosion can be measured directly using close-range stereopairs. Welch *et al.* (1984) report on a method of registering digital data obtained from the stereopairs from two dates to measure the soil loss. This method can provide the spatial data necessary to develop and calibrate physically based erosion models. These data are predominantly obtained from ground-based systems that can permit millimetre accuracies to be obtained.

9.5 WATER QUALITY MODELS

Khorram (1985) used Landsat data and samples taken at 50 sites in San Francisco Bay to develop models of water quality. The water quality parameters studied were salinity, turbidity, suspended solids and chlorophyll *a*. Khorram developed regression equations between the water quality parameters and the mean radiance value of different Landsat bands. His models took the following forms:

Salinity where Y_{ec} is salinity (parts per thousand)

$$Y_{ec} = 91.8 - 19.7 \, (\ln X5) - 11.8 \, (\ln X6) \qquad (9.3)$$

Turbidity where Y_t is turbidity (nephleometric turbidity units)

$$Y_t = 3.70 - 0.40(X4)^2 + 0.8(X5)^2 + 0.09(X6)^2 - 0.57(X7)^2 \quad (9.4)$$

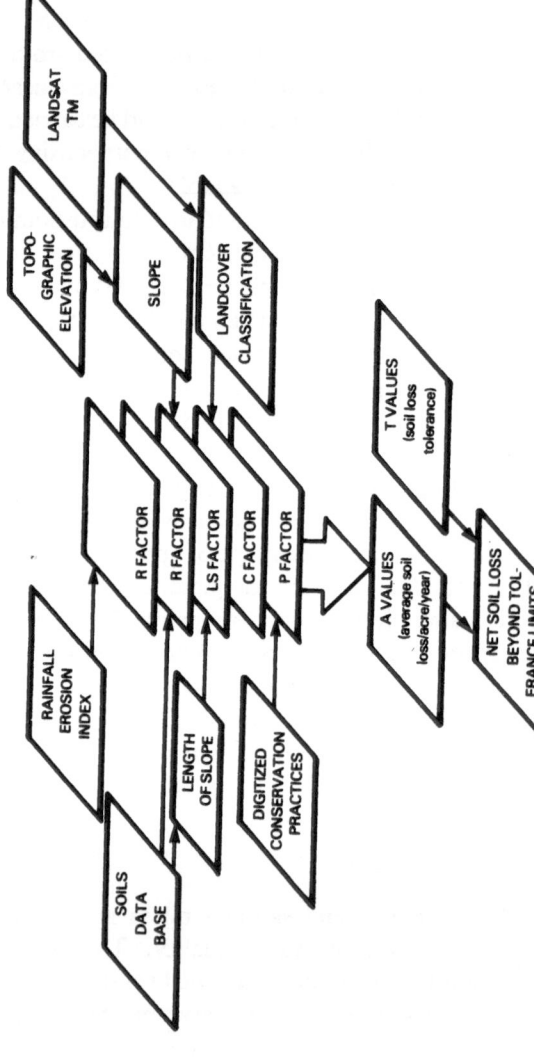

Figure 9.5 A schematic of the universal soil-loss equation database as managed by the GIS (after Pelletier 1985).

Suspended solids where Y_{ss} is total suspended solids (mg l^{-1})

$$Y_{ss} = -79.5 + 24.4 \, (\ln X5) + 0.34 \, (\ln X7) \qquad (9.5)$$

Chlorophyll where Y_{ch} is the chlorophyll concentration (g l^{-1})

$$Y_{ch} = -40.99 + 2.73(X4) - 2.62(X5) + 1.93(X6) + 4.14(X7)$$
$$0.07(X4)^2 + 0.30(X5)^2 - 0.25(X6)^2 - 0.19(X7)^2 \qquad (9.6)$$

where the mean radiance values are $X4$ for Landsat band 4, $X5$ for band 5, $X6$ for band 6 and $X7$ for band 7.

Khorram's (1985) models are generally of the same form reported by other investigators (Rogers *et al.* 1976; Scarpace *et al.* 1978; Shih and Gervin 1980). Khorram made the observation that in the salinity model what may actually be observed is a surrogate of salinity, such as turbidity. In the San Francisco Bay area, the saline water is generally more turbid than fresh-water inflows, and there was thus a strong correlation between salinity and turbidity. This is likely for many of the empirical relationships have have been developed for specific water bodies. A second study by Khorram and Cheshire (1985) illustrates the point that these equations are empirical and applicable only to the water body from which they were derived. Compare equation (9.3) with equation (9.7) below:

$$Y_{ec} = 38.52 - 120.86 \, [X6/(X4 + X5)] \qquad (9.7)$$

Both equations are salinity models, but equation (9.7) was developed for the Neuse River estuary in North Carolina.

More recent work by Khorram *et al.* (1987) has refined the chlorophyll *a* relationship by suggesting different wavelength ratios for different concentration levels. They found a blue–green/near-infrared II (450–520/910–1050 nm) ratio best for relatively low concentrations of chlorophyll *a* and a red/infrared I (630–690/690–750 nm) ratio better for higher concentrations. Plate 8 illustrates the predicted spatial patterns of chlorophyll *a* for the northern portion of San Francisco Bay. This work also tends to emphasize the very useful but site-specific information that can be obtained from remote sensing.

Rogers *et al.* (1976), in their study of Saginaw Bay, emphasize that some of the water quality parameters 'do correlate secondarily with color or volume reflectance to the extent that they all characterize the same water masses'. As an example, their model for chloride, which cannot be measured directly, takes the form

$$Y_{ch} = 99.27 - 57.138(X4/X5) \qquad (9.8)$$

where Y_{ch} is the chloride concentration (mg l^{-1}) and $X4$ and $X5$ are as defined above.

Turbidity and algal pigment were successfully modelled for inland lakes in south-east Australia (Carpenter and Carpenter 1983). Date-independent models were developed by using Landsat data from seven overpasses by including a sun angle correction term. A time correction was also included to compensate for diurnal variations in pigment fluorescence that occurred at the time of ground truth sampling. For illustrative purposes, their equations took the following forms:

$$\log T = 4.51 + (0.304)\ X4 - (0.0727)\ X5 + (0.0534)\ X6 - (10.5)\ SUN$$

$$(9.9)$$

and

$$\log P = 5.48 - 0.114X4 - 0.0546X5 + 5.12SUN - 0.00479TIME$$

$$(9.10)$$

where $T = \ln$ (turbidity) in NTU, $P = \ln$ (pigment) in mg m^{-1}, SUN is the sine of the sun elevation (range 17–47°) at the time of the Landsat overpass, and $TIME$ is the time of day of sample collection (range 0820–1330 h).

Vector analysis of AVHRR and CZCS (coastal zone colour scanner) data has been used to distinguish differences between water colour and pigment concentrations from turbidity (Stumpf and Tyler 1988). The model is based on an empirical relationship between the vector orientation and measured chlorophyll concentration. After making a simple atmospheric correction, the location of algal blooms in the Chesapeake Bay could be delineated (Plate 9). The results indicate that blooms can be identified where the reflectance is between 0.01 and 0.07, and it may be possible to estimate the concentrations of chlorophyll greater than 5 g l^{-1}.

In an analogous study, Dwivedi and Narain (1987) used atmospherically corrected TM bands 1 and 2 to develop a relationship for estimating the concentration of phytoplankton pigment. With their regression equation they were able to map the concentration of the pigment for an area of the Arabian Sea off Azhikal on the west coast of India.

9.6 TROPHIC STATE OF WATER BODIES

The trophic state of a water body is the description of its nutrient status or its productivity. Many of the water quality parameters discussed above are used to characterize a water body's trophic state; however, most of the interest focuses on the nutrients nitrogen and phosphorus. Witzig and Whitehurst (1981) discuss a number of trophic state indices and their estimation from Landsat data. Most trophic state indices are complex combinations of the many individual water quality parameters. In addition, there is no 'standard' index that is applicable for all physico-climatic regions and all types and sizes of lakes. Since many of the trophic state index variables are highly correlated with each other, the use of MSS data should provide a useful synoptic tool for monitoring water bodies.

Wezernak *et al.* (1976) proposed two models related to ratios of specific reflectance bands. They related chlorophyll *a* concentrations to the ratio of a 0.62–0.70 μm band to a 0.42–0.50 μm band, and Secchi disk transparency to the ratio of a 0.50–0.54 μm band to a 0.62–0.70 μm band.

Thermal infrared data have been successfully used to assess lake water quality. A study of Lake Utah (Miller and Rango 1984) correlated HCMM (Heat Capacity Mapping Mission) data with algal concentrations. They found high positive correlations between emitted thermal energy and algal concentrations. At night the opposite occurred and the thermal data exhibited a strong negative correlation with algal concentrations.

9.7   THERMAL POLLUTION

Thermal pollution can be defined as artifically raising or lowering the temperature of a water body to such a degree that life is threatened, either directly or indirectly. Direct thermal pollution changes the temperature so that existing aquatic biota can no longer live and reproduce, and in drastic cases results in outright destruction. Indirect thermal pollution results in changing some other water quality, such as dissolved oxygen, to a deleterious level. Infrared radiometers flown on aircraft have been used on a routine basis for surveying the surface temperatures of water bodies. These instruments can achieve accuracies of 0.5°C over a range of -50 to $+50$°C. There are also a number of techniques that have been developed to generate two-dimensional thermal maps of the measured area.

Thermal maps of surface temperatures have been very useful for siting discharge pipes from factories and thermoelectric generating stations. Diffuser design can be verified by making a series of thermal maps under different flow, weather and effluent discharge conditions. A limitation of thermal remote sensing data is similar to the other applications discussed previously: only surface and near-surface temperatures can be measured with thermal infrared systems. However, deeper temperatures can often be inferred from the surface distribution of temperatures. Temperature maps can provide excellent information for locating depth sampling instrumentation and for designing boat sampling schemes.

9.8   FUTURE APPLICATIONS

There are numerous other applications of remote sensing to water quality problems and measurement. Each of these are unique in that they are site specific. Field data are necessary to develop quantitative relationships that are purely empirical. However, they do provide a large-scale view of the water body that in many cases cannot be obtained by any other manner. Also, the temporal nature of satellite data from differing dates allows the scientist to infer changes in the water body and monitor programmes. The remote sensing approach has one additional feature that should be of interest to workers in water quality. Large-scale imagery or computer maps of reflectance or surface temperature are useful tools for planning *in situ* sampling programmes and for locating sampling buoys and other permanently placed monitoring instrumentation.

Future uses of remote sensing for water quality applications should increase as instruments with more specific spectral bands and improved spatial resolution become available. These advances coupled with basic research into the understanding of water quality and its remote sensing response, and continued improvement of computerized data analysis, should bring about these applications rapidly.

REFERENCES

Abiodun, A. A. (1976) Satellite survey of particulate distribution patterns in Lake Kainji. *Remote Sensing Environ.* **5**, 109–23.

Abiodun, A. A. and Adeniji, H. (1978) Movement of water columns in Lake Kainji. *Remote Sensing Environ.* **7**, 227–34.

Aranuvachapum, S. and LeBlond, P. H. (1981) Turbidity of coastal water determined from Landsat. *Remote Sensing Environ.* **11**, 113–32.

Brakel, W. H. (1984) Seasonal dynamics of suspended-sediment plumes from the Tana and Sabaki Rivers, Kenya: Analysis of Landsat imagery. *Remote Sensing Environ.* **16**, 165–73.

Campbell, W. J. (1981) An application of Landsat and computer technology to potential water pollution from soil erosion. *Satellite Hydrology, 5th Annu. William T. Pecora Symposium on Remote Sensing*, American Water Resources Association, Minneapolis, MN, pp. 616–21.

Carpenter, D. J. and Carpenter, S. M. (1983) Modeling inland water quality using Landsat data. *Remote Sensing Environ.* **13**, 345–52.

Dwivedi, R. M. and Narain, A. (1987) Remote sensing of phytoplankton — An attempt from the Landsat thematic mapper. *Remote Sensing Lett. Int. J. Remote Sensing* **8**, 1563–7.

Feldman, G., Clark, D. and Halpern, D. (1984) Satellite color observations of the phytoplankton distribution in the Eastern Equatorial Pacific during 1982–1983 El Nino. *Science* **226**, 1069–71.

Khorram, S. (1985) Development of water quality models applicable throughout the entire San Francisco Bay and delta. *Photogram. Eng. Remote Sensing* **51**, 53–62.

Khorram, S., Catts, G. P., Cloern, J. E. and Knight, A. W. (1987) Modeling of estuarine chlorophyll *a* from an airborne scanner. *IEEE Trans. Geosci. Remote Sensing* **GE-25**, 662–9.

Khorram, S. and Cheshire, H. M. (1985) Remote sensing of water quality in the Neuse River estuary, North Carolina. *Photogram. Eng. Remote Sensing* 51, 329–41.

Miller, W. and Rango, A. (1984) Using Heat Capacity Mapping Mission (HCMM) data to assess lake water quality. *Water Res. Bull.* 20, 493–501.

Mueksch, M. C. (1984) A splash and sheet erosion model from Landsat data. *Proc. IGARSS'84 Symp., Strasbourg*, ESA SP-215, pp. 295–9.

Munday, J. C., Jr and Alfoldi, T. T. (1979) Landsat test of diffuse reflectance models for aquatic suspended solids measurements. *Remote Sensing Environ.* 9, 169–93.

OKI (Ohio–Kentucky–Indiana Regional Council of Governments) (1975) *A Method for Assessing Rural Non-Point Sources and Its Application in Water Quality Management*, WH-554, Water Resources Planning Division, EPA, Washington, DC.

Pelletier, R. E. (1985) Evaluating non-point pollution using remotely sensed data in soil erosion models. *J. Soil Water Conserv.* 40, 332–5.

Ragan, R. M. and Rogers, R. H. (1978) Use of Landsat satellite remote sensing for regional environmental planning and management. *XV Convention Pan American Federal of Engineering Societies, Santiago, Chile.*

Reid, G. K. (1965) *Ecology of Inland Waters and Estuaries*, 4th edn, Reinhold, New York.

Ritchie, J. C., Schiebe, F. R. and Cooper, C. M. (1984) Use of Landsat TM data to monitor suspended sediments in agricultural impoundments. *Proc. 3rd Australasian Remote Sensing Conf., Queensland, Australia*, Organizing Committee Landsat 84, Brisbane, Australia, pp. 79–87.

Ritchie, J. C., Schiebe, F. R. and McHenry, J. R. (1976). Remote sensing of suspended sediments in surface water. *Photogram. Eng. Remote Sensing* 42, 1539–45.

Rogers, R. H., Shah, N. J. and McKeon, J. B. (1976) Computer mapping of water quality in Saginaw Bay with Landsat digital data. *Proc. ASP 42nd Annu. Meet., Washington, DC*, pp. 584–96.

Scarpace, F. L., Holmquist, K. and Fisher, L. T. (1978) Landsat analysis of lake quality for a state wide lake classification program. *Proc. ASP 44th Annu. Meet.. Washington, DC*, pp. 173–95.

Schecter, R. N. (1976) Resource inventory using Landsat data for area wide water quality planning. *Proc. Symp. on Machine Processing of Remotely Sensed Data, Laboratory for Applications of Remote Sensing, Purdue University, West Lafayette, IN.*

Shih, S. F. and Gervin, J. C. (1980) Ridge regression techniques applied to Landsat investigations of water quality in Lake Okeechobee. *Water Resour. Bull.* 16, 790–6.

Stephens, P. R., MacMillan, J. K., Daigle, J. L. and Cihlar, J. (1985) Estimating universal soil loss equation factor values with aerial photography. *J. Soil Water Conserv.* 40, 293–6.

Stumpf, R. P. and Tyler, M. A. (1988) Satellite detection of bloom and pigment distributions in estuaries. *Remote Sensing Environ.* 24, 385–404.

Thomas, I. L. (1980) Suspended sediment dynamics from repetitive Landsat data. *Int. J. Remote Sensing* 1(3), 285–92.

Viollier, M., Deschamps, P. Y. and Lecomte, P. (1978) Airborne remote sensing of chlorophyll content under cloudy sky as applied to the tropical waters in the Gulf of Guinea. *Remote Sensing Environ.* 7, 235–48.

Welch, R., Jordan, T. R. and Thomas, A. W. (1984) A photographic technique for measuring soil erosion. *J. Soil Water Conserv.* **39**, 191–4.

Wezernak, C. T., Tanis, F. J. and Bajza, C. A. (1976) Trophic state of inland lakes. *Remote Sensing Environ.* **5**, 147–65.

Whitlock, C. H., Witte, W. G., Talay, T. A., Morris, W. D., Usry, J. W. and Pool, L. R. (1981) Research for reliable quantification of water sediment concentration from multispectral scanner remote sensing data. *AgRISTARS Rep.* CP-Z1-04078 9NASA/JSC 17134.

Wischmeier, W. H. and Smith, D. D. (1978) Predicting rainfall erosion losses. *Agricultural Handbook No.* 537, USDA Science and Education Administration.

Witzig, A. S. and Whitehurst, C. A. (1981) Current use of Landsat MSS data for lake trophic classification. *Water Resour. Bull.* **17**, 962–70.

Yarger, H. L., McCauley, J. R., James, G. W. and Magnuson, L. M. (1974) Quantitative water quality with ERTS-1. *Proc. 3rd Earth Resources Technology Satellite Symp.*, NASA SP-351, pp. 1637–51.

10

Water resources management and monitoring

10.1 INTRODUCTION

Hydrology is the scientific study of water, its measurement on the surface of the Earth and its effects. Water and changes in the hydrologic cycle have important economic effects for water delivery and drainage to agriculture and urban areas and for flood control. This larger topic includes economic and social aspects as well as hydrologic ones and is generally studied as water resources management. This chapter is concerned with the application of remotely sensed data specifically to this problem.

There are a number of areas in water resources monitoring and management that can benefit from remotely sensed data. Estimating flood damage and monitoring changes in lake and reservoir volumes are suited to analysis using these remotely sensed data. The temporal nature of Landsat and other satellite data make the information very useful for monitoring changes. Delineation of floodplains and studies of basin morphology and wetlands are other applications suited to using remotely sensed data. In this case it is the availability of large-scale imagery or imagery for areas that are generally inaccessible that are the desirable features of using these data.

10.2 GENERAL APPROACH

The use of remotely sensed data for water resources monitoring and management is basically for mapping. The problem is usually one of identifying a land–water boundary or delineating geological and geomorphic characteristics of an area or determining land use with respect to consumptive use of water. Both analysis of imagery and digital classification techniques have been used successfully. Delineation of land–water boundaries depends on the relative spectral characteristics of soil, vegetation and water (see Figure 2.4). The very low reflectance of water in the near-infrared region of the spectrum makes this waveband the obvious choice for identifying and measuring surface water. Microwave systems are also ideal for identifying land–water boundaries because the dielectric constant for water is considerably greater than that for soil or vegetation-covered soil. Unfortunately, microwave data are not generally available except from experimental satellite or aircraft flights. Land use classification not only involves mapping the area of a given crop or cover but also requires identification of the specific crop or forest species. Multitemporal and multispectral data from Landsat are nearly ideal for this use. In most cases, spectral bands in both the visible (bands 4 and 5) region and at least one of the infrared bands (bands 6 or 7) are needed for effective land use classification. The infrared band often can be used to infer stage of growth and general health of the crop.

10.3 CURRENT APPLICATIONS

Flood monitoring

The area inundated by floods and floodplains can be mapped effectively with remotely sensed data. Satellite data such as those from Landsat can be used to define coverage of an entire river basin but may have some limitations on small basins because of the spatial resolution. Black-and-white photography, infrared photography, thermal infrared data, multispectral scanner data and radar have all been used successfully to map the areal extent of flooding. For most approaches using remotely sensed data, determining areas of inundation depends upon measuring reductions in reflectivity caused by standing or flowing water, high soil moisture, moisture-stressed vegetation, and temperature changes. These effects last for some time after inundation and may be detected for up to 2 weeks or longer after the passage of a flood; thus the need to obtain data exactly during the flood peak may not be necessary. A number of studies using Landsat data and infrared photography have been reported by Rango and Salomonson (1974), Williamson (1974), Morrison and Cooley (1973), Hoyer et al. (1973) and Deutsch and Ruggles (1978). A series of papers directed to this subject can be found in American Water Resources Association (1974). Kruus et al. (1981) reviewed flood applications of satellite data with several examples of the Red River of the North flood of July 1975. They illustrated how flood damage may be estimated from multiple images of the flooded area. Anderson (1978) demonstrated how computer-assisted analysis of colour-infrared aerial photography can be used for flood damage assessment. Interactive digital analysis procedures have been used (Rohde *et al.* 1976) to measure flooded areas. Curry (1977) used aerial photography and satellite imagery to trace the origin of flood water and its movement through the basin to help settle court claims related to flood damage.

Cloud cover is frequently a problem in mapping floods with Landsat data because the 18-day coverage may not provide a clear image during or shortly after the flood. The NOAA satellites have an advantage of more frequent coverage over the target area (twice daily). In spite of the coarser spatial resolution (approximately 1100 m versus 80 m for Landsat MSS and 30 m for the thematic mapper), the NOAA satellite thermal infrared sensor has proved effective in measuring areas of flood inundation (Berg *et al.* 1980; Berg *et al.* 1981; Wiesnet *et al.* 1974).

Tappen *et al.* (1983) used the NOAA-6 and NOAA-7 AVHRR (advanced very-high resolution radiometer) data to develop a flood monitoring scheme. There are several differences between the NOAA-N and Landsat multi-spectral scanner systems that must be considered when using the AVHRR data, as shown below:

1. slightly different sensitivity ranges in the visible and near-infrared portions of the spectrum;
2. two to three additional sensors on the AVHRR, one in the middle infrared and two in the thermal infrared;
3. considerable differences in ground resolution (AVHRR ground resolution varies from about 1×1 km^2 at nadir to about 2.5×6.5 km^2 at the outer edges of a scene, whereas Landsat MSS sensors provide 0.056×0.079 km^2 resolution; the TM data can also be used with a resolution of 0.030×0.030 km^2);
4. large differences in sensor view angle ($110.8°$ for the AVHRR and $11.56°$ for Landsat MSS); and
5. differences in the sensitivity of the radiometers and consequent digitization (0–1024 digital numbers (DNs) for AVHRR and 0–127 DNs for the Landsat MSS sensor).

In the case of multispectral data (from aircraft or satellite), the best choice would be to use a spectral band in the near infrared. For Landsat the best choice would be band 7 (0.8–1.1 μm) because there is little or no reflectance of incident radiation from water in this region. The water appears black in a positive print. Sediment-laden water may give a considerably higher reflectance than clear water and result in confusion when delineating the land–water boundary. Landsat data may be a valuable source of data when no ground surveys are available. For example, Miller (1986) reported that Landsat data were used to verify that overflow from a main channel had occurred when it was not evident from aerial photographs.

Measuring flooded areas

There are several techniques available for processing flood scenes to depict the flooded areas. A very effective method uses optical processing of two scenes of different dates, before and during flooding. The extent of flooding can then be depicted by two colour temporal composites. For example, a band 7 Landsat flood scene can be superimposed on a band 7 pre-flood scene, and a composite produced by an additive colour viewer (Deutsch and Ruggles 1974). Plate 10 illustrates this process.

Black-and-white imagery using the band 7 data is also effective in delineating flooded areas. Plate 11 illustrates the situation before and after the 1973 flooding of the Mississippi River. Such imagery can be overlaid on a topographic map using a zoom transfer scope to relate the flood areas to map areas.

Use of digital data enables an automatic classifier to be developed. There are two basic approaches that use pixel-by-pixel analysis to map water. The first method is based on a threshold approach (Williamson 1974) that takes advantage of the strong absorption of near-infrared radiation. Thus, band 7 (0.8–1.1 μm) radiance values can be expected to be very low. The radiance values over water are seldom greater than 0.2 mW cm^{-2}, unless the water contains impurities. This value is considerably less than values over land. Each pixel is scanned by the computer and wherever the radiance is less than some chosen threshold, say 0.2 mW cm^{-2}, the computer identifies that pixel as water. All other pixels are identified as non-water. For a specific application such as this, it does not matter what the other pixels are as long as the water, non-water decision is acceptable. In this form the digital data can be made into maps or map overlays that depict the spatial distribution of water. This method is fast and easy to set up on a computer but it may result in serious misclassification if there are areas adjacent to the flooded area that have similar low reflectances or if the flood waters are heavily laden with sediment. Typical low-reflectance areas would include urban areas, organic soils and some geological formations. Also, sediment-laden water may have radiances considerably larger than that of clear water. In this case the threshold approach may classify these pixels as non-water and seriously confuse the land–water boundary.

The second approach is to develop a classifier from training site data. This approach can help to minimize the misclassification by taking advantage of reflectance data from more than one spectral band. In addition, this approach may be the only reliable way to differentiate sediment-laden water from adjacent land. Tappen et al. (1983) developed a basic relationship between the spectral returns of AVHRR Channels 1 and 2 for water, soil and vegetation using a large number of NOAA-6 scenes from different seasons and geographic regions. A lookup table classifier was then developed based on the analysis of

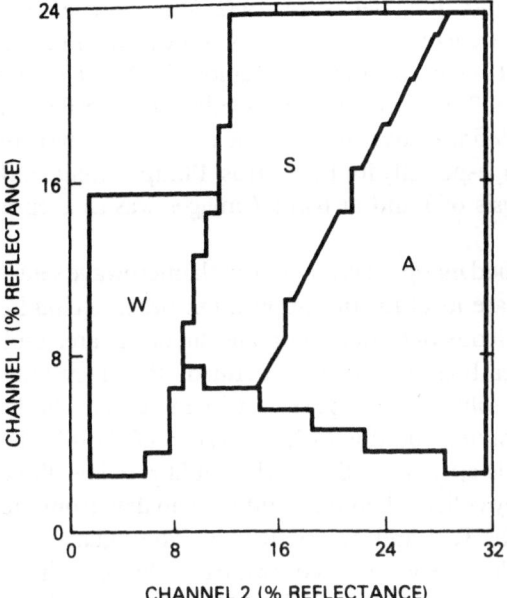

Figure 10.1 An illustration of a stepped-line classifier for identifying water, soil and vegetation; W = water, S = soil and A = vegetation (after Tappen *et al.* 1983).

each channel's reflective relationships for each surface feature. The classifier automatically separated land from water and produced classification maps which were registered to a global coordinate system. Figure 10.1 illustrates this simple classifier that was developed around only three land classes. Testing of the classifier was completed for a number of acquisitions, including coverage of a major flood on the Parana River in Argentina.

Wiesnet and Deutsch (1985) used the Nimbus-7 ocean colour scanner (CZCS) data to analyse the same Parana River flood. They found that the optimum band for drainage delineation was in the blue region (band 5, 0.75 μm). However, a combination of band 3 (0.55 μm) in the green region with band 5 provided easily interpretable images that were superior to either single band. They concluded that in the absence of Landsat data, CZCS imagery provides a very suitable alternative for delineating large floods even though its resolution does not approach that of Landsat but is somewhat better than that of the NOAA AVHRR (825 m for the CZCS as compared with the 1100 m for the NOAA AVHRR). The multiple bands in the visible region can also be used advantageously. As with the AVHRR, one image can usually provide synoptic coverage of most large basins.

In spite of the coarser resolution (900 m versus 80 m for Landsat), the NOAA satellite thermal infrared sensor has proved effective in measuring areas of flood inundation (Berg *et al.* 1981; Wiesnet *et al.* 1974). In addition, the NOAA satellites have the advantage of more frequent coverage (twice daily average versus 18-day coverage for Landsat). Ali *et al.* (1987) have used AVHRR Channel 2 data to delineate flooded areas in Bangladesh. Although there may be a definite advantage for the machine analysis of digital data for flood delineation, especially for large areas, Philipson and Hafker (1981) found that visual analysis of Landsat band 7 images was as accurate as the digital analysis.

All-weather flood mapping is possible with microwave sensors. The active or passive systems are ideal for this application because cloud cover frequently present during floods does not obliterate the target area and in most cases a sharp land–water boundary can be defined. Radar systems are capable of higher spatial resolution than passive systems under similar situations and should be well suited for this task. Lowry *et al.* (1981) demonstrated that airborne synthetic aperture radar (SAR) could provide all-weather flood area delineation. They collected both X- and L-band data from flight lines over the 1979 floods of the Red and Morris Rivers. The radar data were favourably compared with flood extent maps drawn from photographs.

The Shuttle imaging radar (SIR-B) was used to map flood boundaries and assess damage over areas of Bangladesh. The digitally processed and geometrically rectified radar data were compared with earlier Landsat data to estimate the areas of inundation subsidence (Imhoff *et al.* 1987).

Research in this area is still needed to understand how vegetation shadows, depressions, etc. affect the radar response before it can be used on a satellite system (Lowry *et al.* 1981; Bryan 1981). However, a fairly recent study by Ormsby *et al.* (1985) has shown that L-band radar data acquired by Seasat and SIR-A (Shuttle imaging radar-A) can detect flooding under forest canopies. This study indicates that reflection and rescattering actually increases the radar return by 3 to 6 dB. Thus, the longer-wave L-band radar has enhanced return from flooded targets under tree canopies and other targets with relatively large vegetation volumes. The shorter-wavelength X-band data show no enhanced response from tree canopies but do show an enhanced response from smaller vegetation canopies such as marsh grasses. Imhoff *et al.* (1986) demonstrated that water beneath a mangrove forest canopy could be detected at all angles of incidence available from the SIR-B experiments (25.6°, 45.6° and 57.8°). A comparison of the three images is shown in Figure 10.2.

Richards *et al.* (1987) proposed a backscatter model for L-band HH polarized data that explains the increased backscatter observed from flooded forests. It thus appears that radar will be extremely useful for monitoring floods and land–water boundaries under vegetative canopies, but there is still a great deal of research needed to define all the system and target parameters necessary for a useful system.

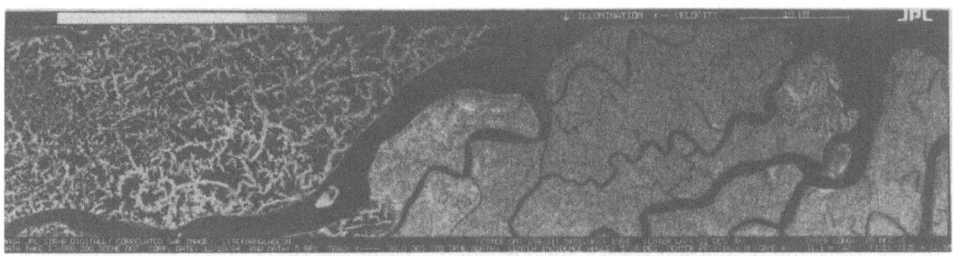

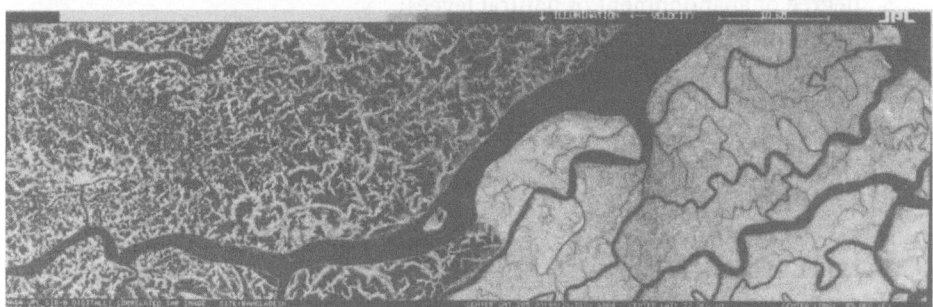

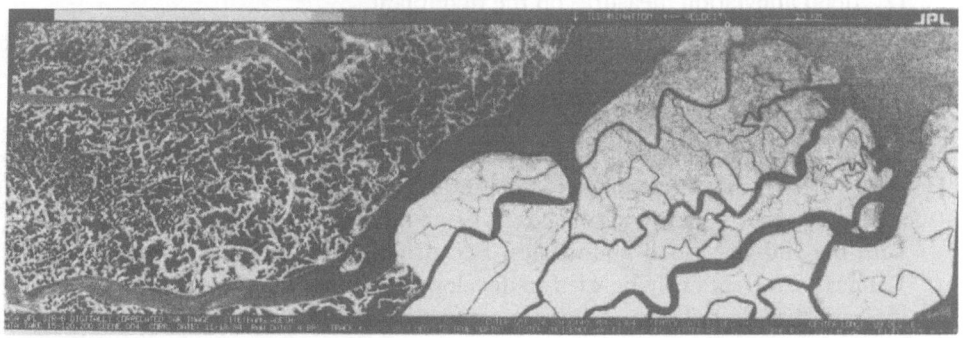

Figure 10.2 Portions of SIR-B synthetic aperture radar images in southern Bangladesh from data takes (top to bottom) 88.2, 104.2 and 120.2, acquired on 11, 12 and 13 October local time. The scenes start in the south (right) at the Sundarbans mangrove forests at the Bay of Bengal and progress north-eastwards to the densely populated agricultural areas south of the Ganges River. Open water areas such as the Bay of Bengal, the Bola River (image centre) and the flood-irrigated fields appear as dark areas. Bright areas in the water can be seen in data take 120.2, which is caused by storm-induced turbulence at the time of overpass. The brighter, more homogeneous-appearing patterns in the south (image right) are the mangrove forests of the Sundarbans. The Sundarbans provide valuable wood and paper pulp resources for Bangladesh. The Supoti test area is outlined in black (Imhoff *et al.* 1986).

Floodplain mapping

Floodplains have been delineated using remotely sensed data to infer the
extent of the floodplain from vegetation changes, soils or some other cultural
features commonly associated with floodplains. Rango and Anderson (1974)
developed the list of indicators shown below that can be used to infer
floodplains from Landsat data:

1. upland physiography;
2. watershed characteristics such as shape, drainage density, etc.;
3. degree of abandonment of natural levees;
4. occurrence of stabilized sand dunes on river terraces;
5. channel configuration and fluvial geomorphic characteristics;
6. backswamp areas;
7. soil moisture availability which could also be a short-time indicator of
   flood susceptibility;
8. soil differences;
9. vegetation differences;
10. land use boundaries;
11. agricultural development;
12. flood alleviation measures on the floodplain.

In a more recent study, Sollers *et al.* (1978) examined multispectral aircraft
and satellite classifications of land cover features indicative of floodplain areas.
They concluded that the satellite data can be used to delineate floodprone areas
in agricultural and limited development areas but may not give good results in
areas with a heavy forest canopy. Fortuitously, Landsat-delineated floodprone
area signatures were found to have, as yet, a relatively unexplained correlation
with the 100-year flood boundaries. Because of this, Sollers *et al.* stated that
Landsat can frequently be used to develop preliminary floodprone area maps
in areas that do not have adequate or up-to-date surveys. The Landsat data
may also be used for checking existing surveys and to identify areas of apparent
discrepancy that merit further detailed ground surveys. In addition, the
Landsat data have the potential for monitoring adherence to development or
land use zoning restrictions in existing floodprone areas. The limitations on the
use of Landsat data are primarily based on the 80 m spatial resolution. Sollers
et al. described how flood and floodprone maps have been produced at
1 : 24 000 and 1 : 62 500 scales using digital Landsat data. Better results and
more accurate delineation of floodprone areas with the TM and SPOT data can
reasonably be expected, although for many legal requirements it is necessary to
map floodprone areas from high-resolution aerial photography.

Wetlands

The environmental importance of wetlands has resulted in an increased
awareness of this resource. A primary requisite to management and protection

of these areas is an accurate map and inventory. Any inventory necessarily includes the area and volume of the water body. Additionally, the inventory should include the boundaries, vegetation types and patterns and water movement directions (Reinhold and Linthurst 1975). Remote sensing is a very useful tool to carry out this type of inventory because wetland studies are generally conducted for large and relatively inaccessible areas.

Low-altitude black-and-white, panchromatic, colour, infrared and thermal infrared photography have all been used successfully in wetlands mapping. Each of these have particular characteristics that can be used advantageously for wetland studies. Colour infrared has been generally preferred for vegetation discrimination, but seasonal changes can confuse the interpretation. In a study of the Great Dismal Swamp, Carter *et al.* (1977) analysed the imagery taken in the dormant season to delineate land–water boundaries, drainage patterns, location of conifers, classification of understorey vegetation and total area of surface water. Imagery obtained in the growing season was used to identify deciduous vegetation.

The resolution of current satellite data limits some of their applications to wetlands. In many cases, Landsat data will not yield as high an accuracy as can be achieved with aerial photography. In a study of Minnesota wetlands, Werth and Meyer (1981) found that aerial photography accuracies were better (approximately 96%) than Landsat accuracies (approximately 71%) when compared with ground data. Gammon *et al.* (1981) showed that Landsat accuracies were not great enough for mapping or management applications in the Great Dismal Swamp. However, in a comparison of SPOT and Landsat MSS data, Ackleson *et al.* (1985) concluded that the higher-resolution SPOT data provided details of small-scale hydrodynamics not identifiable with Landsat data. They also concluded that SPOT data delineated water masses with a high degree of accuracy.

Many inventories are conducted by using vegetation characteristics to delineate the wetland boundaries from air photographs. A very complete summary of this type of analysis has been prepared by Carter (1978) and samples of this type of application have been presented by Carter *et al.* (1977), Morrow and Carter (1978) and Carter *et al.* (1979). Landsat data have been used to make estimates of wetlands water volumes on a monthly basis by combining depth/stage relationships with surface water area (Higer *et al.* 1974). Gervin and Shih (1981) demonstrated how Landsat classification of the complex shoreline marshes of Lake Okeechobee enabled them to correct for a discrepancy in predicting the water volume over a 4-year period. By adjusting the effective surface area according to the vegetation area, they were better able to account for evapotranspiration and improve the total water balance.

An additional aspect of wetlands management involves the extent of dredging, lagoons, drainage and other man-induced changes that have an impact on the natural environment. Remote sensing is well suited to monitor these changes and to make preliminary estimates of the environmental impact.

The temporal aspect of Landsat data allows changes to be observed over time and in some cases predevelopment baseline data to be obtained from early scenes.

Monitoring lake and reservoir volumes

Remote sensing has been used extensively for the delineation of surface water bodies and for the inventory of surface water supplies. Historically, photointerpretation has been used, but recently digital multispectral data have also been used. The photointerpretation approach is time consuming and labour intensive if many water bodies are involved or repetitive inventories are needed. Multispectral data can be used with an automated analysis procedure to achieve rapid results that in many cases also meets the accuracy requirements. The principal advantage of high-altitude or spacecraft sensors is the ability to monitor and inventory water bodies synoptically in areas or situations that are difficult to monitor using conventional methods.

There are a number of morphometric parameters that can be derived from remotely sensed data to describe lakes and reservoirs qualitatively and quantitatively. For extremely accurate work, aerial photographs provide the best sources of data. Satellite data also provide good sources for determining morphometric parameters if the resolution is suitable for the specific use. Some of the parameters obtainable from remote sensing products are: maximum length and maximum effective length; maximum width and mean width; direction of major axis; area of water surface for different elevations; and shoreline length and development of shoreline. Readers are referred to Welch (1948) or Hutchinson (1957) for complete definitions of these parameters.

The remote sensing approach is based on identifying the water and non-water areas in a scene. From a remote sensing perspective, water has a relatively low reflectance especially in the near-infrared portion of the electromagnetic spectrum. Thus, a spectral band in the region from 0.7 to 3.0 μm should be chosen. The Landsat MSS band 7 (0.8–1.1 μm) is the best choice among those available from the Landsat MSS data for locating water bodies.

In general, the accuracy of detection and measuring water bodies has been largely a function of proper identification of water and the sensor spatial resolution. Identification problems involve confusion with areas with a similar appearance, such as cloud shadows, dark soils and urban areas. This problem is most prevalent when using a single spectral band from a multispectral sensor. Aerial photograph interpretation can be used to minimize these errors and to check the initial results. If the data are available, the ratio of visible and near-infrared spectral bands can be helpful for separating water from other surface features. Generally speaking, the low reflectance of the water in both bands will help to separate urban areas, fields, and sometimes cloud shadows and ambiguities produced by variations in atmospheric transmission.

Remote sensing capabilities have been used by the United States Army Corps of Engineers to satisfy statutory requirements to map water impoundments. The objectives of surface water inventories are generally to determine the volume of water being stored in lakes or reservoirs, and then to monitor their changes with time. Remote sensing, for the most part, can only determine the surface of the water and cannot measure the volume directly. Mapping surface water area in reservoirs can be used to estimate the volume of water in storage. The procedures that are used to estimate lake volumes depend on an empirical relationship between surface area or shoreline length and volume. Either an area/volume relationship may be developed or topographic features can be used to estimate the water stage in the reservoir and then relate the stage to water volume. Government agencies can use this approach not only to supplement data they obtain for the reservoirs they manage, but also to maintain an awareness of reservoirs they do not control but which may affect their own management strategies under extreme conditions such as major flooding. These techniques clearly demand high spatial resolution data except under extreme flood conditions.

Water bodies can easily and accurately be mapped with photointerpretation techniques. This approach becomes time consuming and labour intensive if many water bodies are being studied or if repetitive inventories are needed. Use of satellite or other multispectral data has the potential for automating this procedure. Although fast and labour efficient, the accuracy of the multispectral approach is limited by the spatial resolution of the individual pixel. If a computer classification approach is used to measure the area of water bodies and only pure water pixels are counted in the process, the area will also be consistently underestimated.

A mixed pixel or partial pixel approach can be and has been successfully applied in several instances to measure the surface water area. A description of the fundamental concepts behind one partial pixel approach is documented by Malila and Nalepka (1974). This approach works particularly well in cases where water bodies are surrounded by dense vegetation. The vegetation has a high reflectance in the near infrared and the water has a low reflectance. Therefore, where a pixel reflectance falls between these two extremes, the proportion of water and vegetation creating this reflectance can be estimated and converted to appropriate area estimates. For a specific accuracy, it is estimated in this study that when using a mixed pixel approach, the areas of surface water for Landsat data can be measured to ± 1% for water areas of 500 ha and larger, and to ± 8% for water bodies of 5 ha and larger.

The use of TM data with its 30 m nominal resolution will increase these accuracies. The data from the French satellite, SPOT, will be even better, with the expected accuracy being improved to about one-quarter of that shown above.

It is often advantageous to develop a hypsographic curve to relate surface area or shoreline distance to the volume. This often becomes an easy job for man-made reservoirs because use can be made of preconstruction contour maps to establish area–volume relationships. This is not so straightforward for natural lakes. In this case, empirical hypsographic curves must be developed based on lake depth soundings, historical records, or estimates based on an assumed shape.

There are other cases where numerous data are available for some lakes in an area and the lakes all have the same general shape. When this is the case, a simple statistical model can be used. Best and Moore (1981) derived regression equations relating depth and volume to the surface geometry. These equations are listed below for illustrative purposes only and readers must be aware that they are not general equations but are valid only for the area from which they were derived:

$$\text{Depth (cm)} = 0.139(L) - 1.758(A) + 15.059 \qquad (10.1)$$

$$\text{Volume (ha m)} = 0.397(A)^{1.25} \qquad (10.2)$$

where L is the length of the lake in metres and A is the area of the lake in hectares. This approach worked for this study only because there were a large number of lakes and they were all formed the same way in glacial moraine deposits.

Irrigated lands assessment

Remote sensing is a valuable source of data for periodically assessing existing and future water requirements of irrigated crop lands. Proper water resource management is of major importance to the economics of arid and semi-arid regions. In those geographic regions where almost all water used by crops is applied by irrigation, knowing where and how much water is being used is imperative for the efficient operation of water distribution systems. For example, in California agriculture is a major industry and accounts for about 85% of that state's water use (Bauer et al. 1983). However, knowledge of irrigation water use is not limited to arid and semi-arid areas. The largest water use in the Suwannee River Water Management District (Florida) is irrigation (Webster et al. 1981). In these areas, the water use by crops must be estimated. In general, the approach is to identify crop type and estimate the crop acreage from the analysis of aerial photographs or Landsat data.

Aerial photography provides detailed dimensions of field boundaries but may not be useful in estimating crop type. Furthermore, aerial photography may be expensive if large-area coverage is needed or multiple dates are required. Interpretation of aerial photographs is often done manually which can be labour intensive and expensive when the analysis of large areas is required. The Landsat data offer an attractive alternative and are almost ideally suited to this application. The characteristics of large-area and frequent coverage and the capability for computer analysis far outweigh the disadvantage of comparatively poor spatial resolution. In some cases both aerial photography and satellite imagery are combined (Tinney *et al.* 1981a). In addition, the availability of several spectral bands in the visible and near-infrared enable procedures to be developed for identifying crop types.

The general procedure used to estimate water use by irrigated crops is to determine the acreage of different crop types first and then to multiply this area by the water use requirement of that crop. For example, Nellis (1984) used colour-infrared photography and thermal infrared imagery to determine crop type and method of irrigation in central Oregon. This information was then used with a crop consumptive use value and an irrigation efficiency (based on method of application) to determine the water resource use. For arid and semi-arid areas, the water use requirements have been well established for most crops (see, for example, Erie *et al.* (1965)). These data are usually presented as the total water required to produce the crop as depth per unit area. Where supplemental irrigation is used, it is more difficult to determine the water use by crops. If an irrigation district provides the water, their records will be helpful for determining the total water use and for verifying calculations of total water use. In other cases, pumping records, power use, or interviews with the growers may be the only way to estimate the water use.

Crop acreage is easily determined by digital Landsat analysis. In most cases it is desirable to work with some form of ground data. These can be in the form of field surveys, aerial photographs or governmental inventories. The training sites should be selected to represent the variety of crops and farming practices that are to be identified. That is, if an area is being analysed where both irrigated and non-irrigated wheat are being grown, both areas should be represented as training sites. There are a number of automatic classifiers that can be chosen depending upon the objectives of the inventory. Classification accuracies can be quite good but are dependent upon the average field size and the variety of crops grown. In a study mapping irrigated fields in western Kansas, Williams and Poracsky (1981) reported a mapping accuracy of 85–99% depending on the crop (wheat, corn and sorghum) and an aggregated areal statistics accuracy of 99%. In a similar study under more confusing conditions (smaller fields and a more humid climate), Webster *et al.* (1981) overestimated irrigated rye and corn acreages by 32% and 21%, respectively.

The multispectral scanner offers a choice of spectral bands to use for analysis. Most studies have used at least one infrared band in combination with one or more of the visible bands, because healthy well-watered vegetation shows up so well in the infrared data. Some studies have used false-colour imagery created from bands 4, 5 and 7, whereas other studies have used various ratios of visible and near-infrared bands. Webster *et al.* (1981) used false-colour composites of bands 4, 5 and 7 because irrigated crops displayed a much higher reflectance in the near-infrared than a non-irrigated crop of the same type and growth stage.

Multitemporal data are often necessary to monitor crops in climates where several crops can be grown and to identify crop types. Tinney *et al.* (1981b) used Landsat imagery from spring, summer and autumn to account accurately for the large variety of different crops grown in Kern County, California. In humid areas the multitemporal data may be the only means of separating irrigated from non-irrigated acreage. When adequate rain has fallen, there may be little spectral difference between the irrigated and non-irrigated crops. However, the two areas will have different spectral responses if some of the crops are under moisture stress with the irrigated fields exhibiting a much brighter infrared response.

Jensen and Cherry (1980) carried this approach one step further by modelling the water balance of a small (3872 acre) basin in the coastal plain of southern Georgia (USA). The water balance model was built around the modified Penman (1948) evapotranspiration relationship. Inputs to the model were measured rainfall and meteorological data and land use determined from the analysis of Landsat data. Six land use classes were separated using a pattern recognition procedure that resulted in an overall classification accuracy of 89%. Landsat bands 4, 5 and 6 were used to create a three-dimensional feature space that demonstrated the separability between classes. The success of this approach and model was evaluated by comparing the model-generated stream runoff with the measured streamflow. With an approach such as this crop water consumption can be predicted and its impact on existing surface or groundwater supplies evaluated.

10.4 FUTURE APPLICATIONS

Remotely sensed data provide an extremely valuable source of information for many water resource management and monitoring tasks. Menenti and Nieuwenhuis (1986) speculate on how remote sensing can be used for water management in intensively developed countries such as The Netherlands, as well as in developing countries. Perhaps in no other area of hydrology and water resources is remote sensing more immediately applicable. This is primarily because the analysis procedures have been quite well developed and accepted; we only need to substitute satellite data for the more familiar map or survey data.

The availability of frequent data on a relatively large scale allows specific water resource problems to be monitored through various seasons and even from year to year. Such monitoring is, for the most part, highly cost effective, especially when the alternative would necessitate field surveys. In relatively inaccessible areas, remote sensing may be the only way to collect the necessary data. Future development of new and improved sensors and data handling procedures will only make the remote sensing approach more attractive and more useful. Undoubtedly, new applications will be developed and perfected as new instruments and new types of data become available.

In the future, new high-resolution narrow-band sensors and all-weather capability with high-resolution SAR will prove to be extremely valuable for water resource applications.

REFERENCES

Ackleson, S. G., Klemas, V., McKim, H. L. and Merry, C. J. (1985) A comparison of SPOT simulator data with Landsat MSS imagery for delineating water masses in Delaware Bay, Broadkill River, and adjacent wetlands, *Photogram. Eng. Remote Sensing* 51, 1123–9.

Ali, A., Quadir, D. A. and Huh, O. K. (1987) Agricultural, hydrologic and oceanographic studies in Bangladesh with NOAA AVHRR data. *Int. J. Remote Sensing, Tech. Note*, 8, 917–25.

American Water Resources Association (1974) Satellite analysis of the 1973 Mississippi River Floods. *Water Resour. Bull.* 10, 1023–96.

Anderson, W. H. (1978) Flood damage assessment using computer-assisted analysis of color infrared photography. *J. Soil Water Conserv.* 33, 2283–6.

Bauer, E. H., Baggett, J. D., Wall, S. L., Thomas, R. W. and Brown, C. E. (1983) Results of an irrigated lands assessment for water management in California. *Int. Geoscience and Remote Sensing Symp., San Francisco, CA*, vol. I, IEEE, pp. TA 1.1 – TA 1.4.

Bausch, W. C. (1988) Improving irrigation scheduling for corn via remote sensing, *ASAE Summer Meet., Rapid City, SD*, Pap. No. 88-4065, pp. 8.

Berg, C. P., McGinnis, D. F. and Forsyth, D. G. (1980) Mapping the 1978 Kentucky River flood from NOAA-5 satellite thermal infrared data. *Tech. Pap., ACSM-ASP Convention*, American Society of Photogrammetry, St Louis, Mi., pp. 106–11.

Berg, C. P., Matson, M. and Wiesnet, D. R. (1981) Assessing the Red River of the North 1978 flooding from NOAA satellite data. *Satellite Hydrology*, American Water Resources Association, Minneapolis, MN, pp. 309–15.

Best, R. G. and Moore, D. G. (1981) Landsat interpretation of Prairie Lakes and wetlands of eastern South Dakota. *Satellite Hydrology*, American Water Resources Association, Minneapolis, MN, pp. 499–506.

Bryan, M. L. (1981) The use of radar imagery for surface water investigations. *Satellite Hydrology*, American Water Resources Association, Minneapolis, MN, pp. 238–51.

Carter, V. (1978) Coastal wetlands: role of remote sensing. *Proc. Symp. on Technical, Environmental, Socioeconomic and Regulatory Aspects of Coastal Zone Management, San Francisco, CA*.

Carter, V., Garrett, M. K., Shima, L. and Gammon, P. (1977) The Great Dismal Swamp: Management of a hydrological resource with the aid of remote sensing. *Water Resour. Bull.* 13, 1–12.

Carter, V., Malone, D. and Burbank, T. H. (1979) Wetland classification and mapping in western Tennessee. *Photogram. Eng. Remote Sensing* **45**, 273–84.

Curry, D. T. (1977) Identifying flood water movement. *Remote Sensing Environ.* **6**, 51–61.

Deutsch, M. and Ruggles, Jr. F. H. (1974) Optical data processing and projected applications of the ERTS-1 imagery covering the 1973 Mississippi River Valley floods. *Water Resour. Bull.* **10**, 1023–39.

Deutsch, M. and Ruggles, Jr, F. H. (1978) Hydrological applications of Landsat imagery used in the study of the 1973 Indus River flood, Pakistan. *Water Resour. Bull.* **14**, 261–74.

Erie, L. J., French, O. F. and Harris, K. (1965) Consumptive use of water by crops in Arizona. *University of Arizona Agricultural Experiment Station, Tech. Bull.* 169.

Gammon, P. T., Rohde, W. G. and Carter, V. (1981) Accuracy evaluation of Landsat digital classification in the Great Dismal Swamp. *Satellite Hydrology*, American Water Resources Association, Minneapolis, MN, pp. 463–73.

Gervin, J. C. and Shih, S. F. (1981) Improvements in lake volume predictions using Landsat data. *Satellite Hydrology*, American Water Resources Association, Minneapolis, MN, pp. 479–84.

Higer, A. L., Coker, A. E. and Cordes, E. H. (1974) Water management models in Florida using ERTS-1 data. *Proc. 3rd Earth Resources Technology Satellite-1 Symp.*, NASA SP-351-V1, pp. 1071–88.

Hoyer, B. E., Hallberg, G. R. and Taranik, J. V. (1973) Seasonal multispectral flood inundation mapping in Iowa. *Management and Utilization of Remote Sensing Data, Proc. Symp. of American Society of Photogrammetry, Sioux Falls, SD*, pp. 130–41.

Hutchinson, G. E. (1957) *A Treatise on Limnology*, Wiley, New York.

Imhoff, M., Story, M., Vermillion, C., Khan, F. and Polcyn, F. (1986) Forest canopy characterization and vegetation penetration assessment with spaceborne radar. *IEEE Trans. Geosci. Remote Sensing* **GE-24**, 535–42.

Imhoff, M. L., Vermillion, C., Story, M. H. Choudhury, A. M., Gafoor, A. and Polcyn, F. (1987) Monsoon flood boundary delineation and damage assessment using space borne imaging radar and Landsat data. *Photogram. Eng. Remote Sensing* **53**, 405–13.

Jensen, J. R. and Cherry, Jr. D. L. (1980) Landsat crop identification for watershed water balance determinations. *American Society of Photogrammetry, 46th Annu. Meet., St. Louis, MO*, pp. 65–80.

Kruus, J., Deutsch, M., Hansen, P. L. and Ferguson, H. L. (1981) Flood applications of satellite imagery. *Satellite Hydrology*, American Water Resources Association, Minneapolis, MN, pp. 292–301.

Lowry, R. T., Langham, E. J. and Mudry, N. (1981) A preliminary analysis of SAR mapping of the Manitoba flood, May 1979. *Satellite Hydrology*, American Water Resources Association, Minneapolis, MN, pp. 316–23.

Malila, W. A. and Nalepke, R. F. (1974) Advanced processing information extraction techniques applied to ERTS-1. *Proc. 3rd Earth Resources Technology Satellite Symp.*, NASA SP-351, pp. 1743–72.

Menenti, M. and Nieuwenhuis, G. J. A. (1986) Remote sensing in the water management practice. *Neth. J. Agric. Sci.* **34**, 317–28.

Miller, S. T. (1986) The Quantification of Floodplain Innundation by the Use of

LANDSAT and Metric Camera Information, Belize, Central America. *Proc. Symp. Remote Sensing for Resources Development and Environmental Management*, Enschede, The Netherlands, pp. 733–8.

Morrison, R. B. and Cooley, M. E. (1973) Assessment of flood damage in Arizona by ERTS-1 imagery. *Proc. Symp. on Significant Results Obtained from ERTS-1, New Carrollton, MD*, vol. 1, NASA, Washington DC, pp. 755–60.

Morrow, J. W. and Carter, V. (1978) Wetland classification on the Alaskan north slope. *5th Canadian Symp. on Remote Sensing, Victoria, British Columbia*.

Nellis, M. D. (1984) A remote sensing approach for modeling water resource use. *Water Resour. Bull.* **20**, 789–93.

Nieuwenhuis, G. J. A. and Bouwmans, J. M. M. (1986) Application of multispectral scanning remote sensing in agricultural water management problems. *Proc. Symp. on Remote Sensing for Resource Development and Water management Problems*, Enschede, The Netherlands, pp. 489–94.

Ormsby, J. P., Blanchard, B. J. and Blanchard, A. J. (1985) Detection of lowland flooding using active microwave systems. *Photogram. Eng. Remote Sensing* **51**, 317–28.

Penman, H. L. (1948) Natural evaporation from open water, bare soil and grass. *Proc. R. Soc.* A **193**, 129–45.

Philipson, W. R. and Hafker, W. R. (1981) Manual vs. digital Landsat analysis for delineating river flooding. *Photogram. Eng. Remote Sensing* **47**, 1351–6.

Rango, A. and Anderson, A. T. (1974) Flood hazard studies in the Mississippi River basin using remote sensing. *Water Resour. Bull.* **10**, 1060–81.

Rango, A. and Salomonson, V. V. (1974) Regional flood mapping from space. *Water Resour. Res.* **10**, 473–84.

Reinhold, R. J. and Linthurst, R. A. (1975) Use of remote sensing for mapping wetlands. *Am. Soc. Civil Eng. J. Transp. Eng.* **101**, TE-2, 189–98.

Richards, J. A., Woodgate, P. W. and Skidmore, A. K. (1987) An explanation of enhanced radar backscattering from flooded forests. *Int. J. Remote Sensing* **8**, 1093–100.

Rohde, W. G., Taranik, J. V. and Nelson, C. A. (1976) Inventory and mapping of flood inundation using interactive digital image analysis techniques. *Proc. 2nd Annu. Pecora Memorial Symp., Sioux Falls, SD*, pp. 131–43.

Shih, S. F. (1981) Use of Landsat data to improve water storage information in conservation area, Florida. *Proc. Int. Symp. on Rainfall–Runoff Applied Modeling in Catchment Hydrology*, Water Resources Publications, Littleton, CO, pp. 511–18.

Sollers, S. C., Rango, A. and Henninger, D. L. (1978) Selecting reconnaissance for flood plain surveys. *Water Resour. Bull.* **14**, 359–73.

Tappen, G., Horvath, N. C., Doraiswamy, P. C., Engman, T. and Goss, D. W. (1983) Use of NOAA-n satellites for land/water discrimination and flood monitoring. *NASA-AgRISTAR Rep.* EW-L3-04394.

Tinney, L., Holloway, J., Baggett, J. and Estes, J. (1981b) A multi-stage mapping approach for an irrigated croplands inventory. *Satellite Hydrology*, American Water Resources Association, Minneapolis, MN, pp. 694–5.

Tinney, L., Wall, S., Colwell, R. and Estes, J. (1981a) Applications of remote sensing for California irrigated lands assessment. *Satellite Hydrology*, American Water Resources Association, Minneapolis, MN, pp. 689–93.

Webster, K. B., Lucas, J. R., Musgrove, R. J. and Higer, A. L. (1981) Selected irrigation acreage estimates in northern Florida from Landsat data. *Satellite Hydrology*, American Water Resources Association, Minneapolis, MN, pp. 701–5.

Welch, P. S. (1948) *Limnological Methods*, McGraw-Hill, New York.

Werth, L. F. and Meyer, M. P. (1981) A comparison of remote sensing techniques for Minnesota wetlands classification. *Satellite Hydrology*, American Water Resources Association, Minneapolis, MN, pp. 492–7.

Wiesnet, D. R. and Deutsch, M. (1985) A new application of Nimbus-7 CZCS: Delineation of the 1983 Parana River flood in South America. *Proc. of the American Society of Photogrammetry, 45th Annu. Meet., Washington, DC*, pp. 746–54.

Wiesnet, D. R., McGinnis, D. F. and Pritchard, J. A. (1974) Mapping of the 1973 Mississippi River floods by the NOAA-2 satellite. *Water Resour. Bull.* **10**, 1040–9.

Williams, T. H. and Poracsky, J. (1981) Mapping irrigated lands in western Kansas from Landsat. *Satellite Hydrology*, American Water Resources Association, Minneapolis, MN, pp. 707–14.

Williamson, A. N. (1974) Mississippi River flood maps from ERTS-1 digital data. *Water Resour. Bull.* **10**, 1050–9.

11

Future developments

11.1 INTRODUCTION

Many techniques for using remotely sensed data have been described in this book. Most of them have been proven by field use and are either in use by an operational agency or under development for such agencies. However, the interest of the scientific community in global change has resulted in an increased awareness of hydrological science as opposed to water resources, and also in several new initiatives that are likely to change remote sensing and hydrology in the future. While it is hard to predict the scientific developments that will occur, it behooves us to look at the likely developments over the next few years.

The existence of abundant water in all three phases (solid, liquid and gas) is one of the most distinctive characteristics separating the Earth from the other planets of the solar system, and the climate is significantly ameliorated on Earth because of changes between these phases. It has always been recognized that water is a key substance in the study of the Earth, of climatology and meteorology, plant ecology or geophysics. However, the difficulty of making measurements and conducting the analyses of complicated systems involving water and the environment, together with the great demand for immediate application of hydrologic knowledge for water supply and irrigation projects, has meant that water resources studies have taken precedence over studies in scientific hydrology. This predominance has been present for many years, and in many countries.

Several factors are changing the comparative lack of attention to the studies of scientific hydrology. First, the obvious changes of atmospheric carbon dioxide and methane concentrations which are being measured have led to pressure for a more quantitative understanding of the Earth system, which requires an understanding of the hydrologic cycle. Second, the computational hardware and software tools needed to handle these complex global problems are being developed to the point where they can carry out the calculations involved. Third, remote sensing instruments and conventional data communications and archiving are developing to the point where the data needed for initializing and updating global Earth system models are available in a calibrated form that allow comparisons to be made over time. These are allowing great changes to be made in the type of hydrologic questions that can be addressed (e.g. Eagleson 1990) and also allowing a much greater emphasis on hydrology studies than previously. The advances in each of the areas discussed earlier will be described in turn.

11.2 NEW SENSORS AND PLATFORMS

Several new satellites are planned for launch over the next decade which will carry payloads and make measurements relevant to the land part of the hydrologic cycle. There are several satellites, such as ERS-1 to be launched by the European Space Agency, or J-ERS-1 to be launched by the Japanese, which have primarily oceanographic applications but which will also be used for acquiring data over land. Both will carry single-polarization single-wavelength synthetic aperture radars, plus radiometers at various wavelengths which may also give useful data over land. The duty cycles and coverage of the SARs will be limited in both cases, but the use of SAR data for mapping, particularly in polar areas where overlapping satellite ground tracks mean that the repeat interval between data on different days is reduced, is clearly of interest to the hydrology community.

A list of satellite sensors that have been launched or are under construction is given in Table 2.1. It will be seen that several new satellites and instruments are of interest to hydrologists, including the SARs mounted on the ERS-1, J-ERS-1 and Radarsat spacecraft. Continuing high spatial resolution data from the Landsat and SPOT satellites, passive microwave data from the special sensor/microwave imager (SS/MI) on the DMSP satellite series, and continuing meteorological satellite coverage from the NOAA, GOES, GMS and Meteosat series all mean that the remotely sensed techniques described in this book can continue to be employed and expanded upon. However, there are many other sensors and satellites being planned that will have considerable hydrologic interest. Table 11.1 describes proposed satellites which are currently in detailed planning.

The Tropical Rainfall Measurement Mission and Eos are two of the planned systems that will be of great interest to hydrologists. The Tropical Rainfall

Table 11.1 Research satellites in the planning phase, but not yet under construction (to be compared with Table 2.1 for more operational or complete satellites)

Tropical Rainfall Measurement Mission (NASA/Japan):
    Objective: estimate tropical rainfall
    Orbit: $+30°$; 320 km, rapid precession
    Launch date: January 1994
    Payload:
        AVHRR: vis. and 10 m IR, 1 km resolution, 600 km swath
        Microwave radiometer: 19, 37, 90 GHz dual polarized, 10 km resolution,
                    600 km swath
        Radar: 14 and 24 GHz, 4 km footprint, 250 m range resolution, 220 km swath

NASA Polar Platforms (NPOP-1 and NPOP-2):
    Objective: Earth observation for global change
    Orbit: polar, 705 km, sun-synchronous
    Launch dates: January 1997, 1999, 2002, 2004, 2007, 2009
    Payload chosen from:

Instruments	Mass (kg)	Power (W)
Optical surface imaging:		
MODIS-N	278	325
MODIS-T	159	203
MISR	113	54
HIRIS	897	375
ITIR	422	878
CERES	125	104
EOSP	25	13
Tropospheric sounding:		
AIRS	120	405
AMSU	184	125
HIMSS	412	89
Radars:		
ALT	306	281
SCANSCAT	704	729
STIKSCAT	330	378
Tropospheric composition:		
MOPITT/TRACER	102	192
TES	675	602
Upper-atmosphere sounding:		
HIRRLS/DLS	221	286
SAGE III	81	17
SAFIRE	422	410
MLS	620	774
SWIRLS	134	265

(*continued*)

Table 11.1   *(continued)*

Particles and fields:		
IPEI	59	8
XIE	121	47
SEM	40	38
GOS	99	104
ENACEOS	42	21
POEMS	181	149
Others:		
LIS	16	20
SOLSTICE	282	10
GLRS	603	608
GGI	130	117
WBDCS	48	14
COM PKG	83	25
Total	8034	7666

European Polar Platform-1 (EPOP-1):
  Objective: earth science research
  Orbit: sun-synchronous, 824 km, 10 : 00 a.m. equator crossing
  Launch date: January 1997
  Payload:

Instruments	*Description*
AMRIR	A NOAA operational instrument on the ESA platform; a follow-on to the AVHRR and HRS may also include characterization of the low-light-level channel flown on the DMEP
AMSU-A	Microwave radiometer instrument
AMSU-B	Microwave radiometer
DB	Direct broadcast communication
ARGOS	Provides platform location and sensor data from fixed and mobile platforms from anywhere in the world
CERES	Two broadband scanning radiometers, one in a cross-track mode and one with rotating scan plane; three channels on each scanner will be a total channel (0.2 to $> 100\,\mu$m), SW channel (0.2–$3.5\,\mu$m) and an LW channel (6–$25\,\mu$m); nadir field of view about 25 km
SARSAT	International cooperative search and rescue system
SCATT-2	An active microwave system intended to measure surface wind speeds (vector) over the oceans in the range from 4 to 20 m s^{-1}
MERIS	Imaging spectrometer designed to operate in very narrow bands in the vis./near-IR band of the spectrum with a resolution tuned to ocean/meteorology/climate applications

SAR-C	C-band SAR
MIMR	A dual-polarization passive imaging microwave radiometer intended to provide information on surface characteristics
ATLID	Four cross-track scanning backscatter lidar
LISA	Infrared interferometer for the investigation of stratospheric composition

European Polar Platform-2 (EPOP-2):
   Objectives: earth science research
   Orbit: sun-synchronous, 705 km, 10 : 00 a.m. equator crossing
   Launch date: July 2000
   Payload:

Instruments	Description
SAR-C	C-band SAR
HRIS	High-resolution imaging spectrometer
HRTIR	Push broom radiometer
MERIS	Imaging spectrometer designed to operate in very narrow bands in the vis./near-IR part of the spectrum with a resolution tuned to ocean/ meteorology/climate applications
ALT-2	An adaptive pulse-limited radar altimeter designed to operate over land and ice as well as ocean; a dual-frequency radar altimeter, including a passive microwave radiometer mode for humidity correction
LISA	Infrared interferometer for the investigation of stratospheric composition

Japanese Polar Platform (JPOP-1)
   Objectives: earth science research
   Orbit: sun-synchronous, 800 km
   Launch date: September 1998
   Payload:

Instruments	Description
LAWS	Lidar atmosphere wind sounder—tropospheric winds
AMSR	Advanced microwave scanning radiometer to measure water vapour, precipitation rate, and snow and ice extent
AVNIR	Advanced visible and near-IR radiometer
OCTS	Ocean colour and temperature sensor
SAR-L	L-band synthetic aperture radar
SAR-X	X-band synthetic aperture radar

Measurement Mission (Simpson *et al.* 1988), to be launched into an orbit inclined at about 30°, will have on board a rainfall radar that will be capable of estimating rainfall rates over land. The orbit is such that any point within the tropics will be sampled twice each month at every hour.

Eos (Butler *et al.* 1988) and its counterpart European and Japanese platforms will lead to considerable advances in the understanding of the whole earth sciences, including hydrology. Although the draft manifest of instruments shown in Table 11.1 is impressive, the organization of the data and of other earth science data in an information system where time series of all the data will be easily available is of at least equal importance. The data system will also allow many types of data to be used simultaneously to calibrate or be assimilated into numerical models.

11.3   ADVANCES IN NUMERICAL MODELLING

All the models described in the first ten chapters of this book have been calibrated or verified by a single type or only a very few types of data. This is because many of the estimation techniques used are inverse techniques, where there are rather fewer measurements than unknowns, and these can easily be mathematically intractable. People are only just starting to understand how to use several types of data together in numerical models (e.g. Hollingsworth *et al.* 1986). Computers have also become larger and are more capable of handling complex problems.

While standard hydrologic models continue to improve and become more physically based, there is a whole new class of model that is now receiving attention from hydrologists (Eagleson 1990). These are global models that include the whole water budget. While it may be recalled from Figure 1.1 and Table 1.1 that almost all the water at the surface of the Earth is in the oceans, locked in ice sheets or in deep groundwater, the cycling of the remaining water between the ocean and land surfaces and the atmosphere regulates the Earth's weather and climate, as well as its hydrology.

Early global models were developments of atmospheric general circulation models for weather forecasting, and included only the simplest representations of atmospheric hydrologic processes such as condensation and only minimal descriptions of the land and ocean surfaces. However, as these models were refined for climatological studies, ocean circulation and the air–sea interaction had to be included. The land surface hydrology has also begun to be represented in global models, with models varying from the very simple (Manabe 1969) representation of the land surface as a 'bucket', to the very complicated (Dickinson *et al.* 1986; Sellers *et al.* 1986).

The level of complexity that ought to be included varies from representation to representation, though it is clearly hard to validate a model with many parameters. The level of complexity should match the level of physical representation elsewhere in a global model: a global model is only as good as the least good physical representation within it. Thus it may be seen from Dickinson and Henderson-Sellers (1988) that more realism in the representation of the land surface in one widely used climate model, the National Centre for Atmospheric Research community climate model, will be made first through the inclusion of improved atmospheric radiation and not through more complete surface models alone. The need for the inclusion of mesoscale atmospheric processes in climate models for estimating rainfall will also lead to a better understanding of how surface processes and large-scale atmospheric circulation are linked, and will help with the estimation of surface variables such as evaporation. The atmospheric sounders such as AIRS on Eos (Table 11.1) will probably be able to calibrate such models for use as estimators of evaporation.

One feature of the large-scale hydrologic models that is not well described anywhere is the spatial variability of the land surface. Some advances are being made (for instance, see Entekhabi and Eagleson (1989)), but many models just average measured variables by taking the arithmetic mean for all quantities over large areas. Remote sensing can help with dealing with the spatial variability of the land surface through observations, but experiments are needed to understand what these remotely sensed data are related to and how they can be used.

11.4 OBSERVATIONAL ADVANCES

There have been several recent experiments to try to understand the spatial variability of hydrologic fluxes at the land surface and to relate these to remotely sensed data. Two in particular will be noted here because of their size. The hydrologic–atmospheric pilot experiment, or HAPEX (Andre *et al.* 1986), took place in south-western France in 1986. A $100 \times 100 \, \text{km}^2$ area was covered with radiosondes to estimate the atmospheric component of the water budget, runoff and soil moisture samples, and some surface evaporation and sensible heat flux measurement sites. Aircraft were also flown over the site to estimate sensible and latent heat fluxes and to make remote sensing measurements. Early measurements have shown that there are considerable variations in the radiant surface temperature in the site (Schmugge and Goutorbe 1989), which is half agriculture and half forested, and that sites which contain different types of land use in a limited area (e.g. trees and grass) behave differently from areas that are of uniform land cover (e.g. all trees or all grass) (Shuttleworth 1988). Early modelling studies have shown, however, that the evaporation and water balance can be modelled as a whole using a coupled mesoscale atmospheric and surface model (Andre *et al.* 1988), leading one to hope that the data requirements for routine monitoring will not be prohibitive. Clearly, more work needs to be done, though, on how to couple surface and mesoscale atmospheric models and then calibrate them with remotely sensed data.

Under the international satellite land surface climatology project (ISLSCP), the first ISLSCP field experiment was carried out on a natural prairie grassland in Kansas in 1987 and 1989 (Sellers *et al.* 1988). It is similar in motivation to HAPEX, but with more intensive surface and remote sensing measurements in a more limited area (15×15 km^2). The experimental team was large (29 principal investigators) and a wide variety of useful results should be forthcoming, but it has already been shown (Shuttleworth *et al.* 1989) that the site appeared to behave very uniformly spatially in 1987, when the vegetation was not frequently lacking water, and that the changes in remotely sensed data, particularly reflectance changes, tracked the seasonal changes in the site rather than any shorter-term changes. Results by Wang *et al.* (1989) related to changes in remotely sensed soil moisture and runoff, which also seemed to relate well on average, have already been discussed in Chapter 6.

If, as it appears, local spatial variability can be assumed random for continental-scale hydrologic purposes, then remote sensing at a global scale becomes possible. The results of Chang *et al.* (1987), related to global mapping of soil moisture, have already been discussed in Chapter 4. Similar global maps have been produced for vegetation using visible and near-infrared AVHRR data (Justice *et al.* 1985) and from 37 GHz microwave polarization difference data from the Nimbus-7 SMMR instrument (Choudhury 1988a). Choudhury (1988b) has shown that the two data sets are closely related and also that the microwave data have a remarkably close ($r^2 = 0.97$) empirical relationship with climatological average evaporation. As the radiative transfer and hydrologic models are refined, so that this relationship can be given a better physical basis, this type of relationship is likely to provide a new and exciting use for remotely sensed data.

11.5 FUTURE CONSIDERATIONS

There are many new and exciting observations of the hydrologic cycle that are going to be available from new satellite systems, and, further, new models to allow the data to be used quantitatively in physical interpretations. As we become more proficient at dealing with continental-scale problems, new uses will also be found for existing data sets. Many experiments will be needed to tie these observations together with conventional observations and validate them, both through field experiments such as FIFE (Sellers *et al.* 1988) and through global experiments such as the global energy and water cycle experiment (GEWEX) (WMO 1988) of the World Climate Research Program. Remote sensing can provide many of the suitable data to supplement the conventional data to expand hydrology in new and exciting directions, and also provide entirely new data types and forms that will help hydrologists tackle heretofore unsolvable questions.

REFERENCES

Andre, J-C., Goutourbe, J-P. and Perrier, A. (1986) HAPEX-MOBILHY: A hydrologic–atmospheric experiment for the study of water budget and evaporation flux at the climatic scale. *Bull. Am. Meteorol. Soc.*, **67**, 138–44.

Andre, J. C. *et al.* (1988) Evaporation over land surfaces: first results from HAPEX-MOBILHY special observing period. *Ann. Geophys.* **6**, 477–92.

Butler, D. *et al.* (1988) *From Pattern to Process: The Strategy of the Earth Observing System*, NASA, Washington, DC.

Chang, A. T. C., Foster, J. L. and Hall, D. K. (1987) Nimbus-7 derived global snow cover parameters. *Ann. Glaciol.* **9**, 39–44.

Choudhury, B. J. (1988a) Microwave vegetation index: A new long-term global data set for biospheric studies. *Int. J. Remote Sensing* **9**, 185–6.

Choudhury, B. J. (1988b) Relating Nimbus-7 37 GHz data to global land surface evaporation, primary productivity and the atmospheric CO_2 concentration. *Int. J. Remote Sensing* **9**, 169–76.

Dickinson, R. E. and Henderson-Sellers, A. (1988) Modelling tropical deforestation: a study of GCM land surface parameterizations, *Q. J. R. Meteorol. Soc.* **114**, 439–62.

Dickinson, R. E., Henderson-Sellers, A., Kennedy, P. J. and Wilson, M. F. (1986) Biosphere–Atmosphere transfer scheme (BATS) for the NCAR community climate model, *NCAR, Boulder, CO, Tech. Note* TN 275 + STR.

Eagleson, P. S. (1990) Global change, a catalyst for the development of hydrologic science. Symposium on global change systems, special session on climate change and hydrology. *Am. Meterorl. Soc.*, Anoheim, CA, pp. 1–3.

Entekhabi, D. and Eagleson, P. S. (1989) Land surface hydrology parameterization for atmospheric general circulation models including subgrid scale spatial variability. *J. Climate* **2**, 816–31.

Hollingsworth, A., Shaw, D. B., Lowenberg, P., Illari, L., Arpi, K. and Simmons, A. J. (1986) Monitoring of analysis and observation quality by a data assimilation system. *Mon. Weather Rev.* **114**, 861–79.

Justice, C. O., Townshend, J. R. G., Holben, B. N. and Tucker, C. J. (1985) Analysis of the phenology of global vegetation using meteorological satellite data. *Int. J. Remote Sensing*, **6**, 1271–318.

Manabe, S. (1969) Climate and ocean circulation: 1. The atmospheric circulation and the hydrology of the Earth's surface. *Mon. Weather Rev.* **97**, 739–74.

Schmugge, T. J. and Goutorbe, J. P. (1989) Remotely sensed surface temperature observations in HAPEX. *Proc. IGARSS'89, IEEE, No.* 89CH2768-0, pp. 2127–9.

Sellers, P. J., Hall, F. G., Asrar, G., Strebel, D. E. and Murphy, R. E. (1988) The first ISLSCP field experiment. *Bull. Am. Meteorol. Soc.* **69**, 22–7.

Sellers, P. J., Mintz, Y., Sud, Y. C. and Dalcher, A. (1986) A simple biosphere model for use within general circulation models. *J. Atmos. Sci.* **43**, 505–31.

Shuttleworth, W. J. (1988) Macrohydrology: The new challenge for process hydrology. *J. Hydrol.* **100**, 31–56.

Shuttleworth, W. J., Gurney, R. J., Hsu, A. Y. and Ormsby, J. P. (1989) FIFE: The variation in energy partition at surface flux sites. *Remote Sensing and Large-Scale Global Processes* (ed. A. Rango), *IAHS Publ.* 186, Wallingford, England, pp. 67–74.

Simpson, J., Adler, R. F. and North, G. R. (1988) A proposed Tropical Rainfall Measuring Mission (TRMM) satellite. *Bull. Am. Meteorol. Soc.* **69**, 278–95.

Wang, J. R., Shiue, J. C., Schmugge, T. J. and Engman, E. T. (1989) Mapping soil moisture with L-band radiometric measurements. *Remote Sensing Environ.* **27**, 305–12.

WMO (1988) *Concept of the Global Energy and Water Cycle Experiment*, World Meteorological Organization, Geneva, WCR P-5, WMO/TD 215.

Index